Library of
Davidson College

A PRIMER ON STRUCTURED PROGRAM DESIGN

A PRIMER ON
Structured Program Design

GARY L. RICHARDSON
Texaco, Inc.

CHARLES W. BUTLER
University of Arkansas

JOHN D. TOMLINSON
Air Force Data Services Center

a petrocelli
book
new york / princeton

Copyright © 1980 Petrocelli Books, Inc.

All rights reserved.
Printed in the United States

1 2 3 4 5 6 7 8 9 10

Designed by Joan Greenfield

Library of Congress Cataloging in Publication Data

Richardson, Gary L
   A primer on structured program design.

   (PBI series for the computing and data processing professional)
   Includes index.
   1. Structured programming. I. Butler, Charles William, 1948-   joint author. II. Tomlinson, John D., 1946-   joint author. III. Title. IV. Series.
   QA76.6R483   001.64'2   79-27789
   ISBN 0-89433-085-3
   ISBN 0-89433-011-8 pbk.

CONTENTS

Preface ix

1. Background 1

    Traditional software development problems 3
    Evolution of structured programming concepts 4
    Structured programming techniques 7
    Review questions 9
    Notes and references 9

2. Evolution of Program Design Methodologies 11

    Early design techniques (before 1960) 15
    Design methodologies of the 1960s 17
    Computer-assisted design techniques 20
    Contemporary design techniques 21
    Future trends 27
    Summary 28
    Review questions 29
    Notes and references 30

3. Top-Down Design 31

    Traditional approach 32
    Top-down approach 32
    Logic design 35
    Top-down design constructs 39
    Summary 41

Table of Contents

Review questions 43
Notes and references 43

4. Structured Programming 45

Logic constructs 46
Structured programming with high-level languages 48
Supplementary coding standards 57
Review questions 59
Notes and references 59

5. Structured Design Techniques 61

System life cycle 63
Functional design concepts 65
Role of HIPO in systems design 68
Local implementation of HIPO 74
Summary 80
Review questions 81
Notes and references 81

6. Pseudo-Coding 83

Syntax structure of pseudo-code 84
Formalizing the syntax 86
Formatting considerations 97
Condition handling 98
Use of comments 99
Philosophy of implementation 101
Format and verifier program 104
A design exercise 105
Summary 111
Problems 111
Case study 112

Supplement: ABC Auto Parts—A Case Study 124

7. Program Style and Debugging 143

Programming style 144

Debugging   151
Debugging checklist   157
Debugging case study   159
Review questions   172
Notes and references   172

8. Data Base Considerations — 175

Internal data base   176
External data base   185
Summary   192
Review questions   193
Notes and references   194

9. Managing the Programming Process — 195

Basic components of programming management   197
Program development cycle   200
Summary   207
Problems   208
Notes and references   208

10. Final Remarks — 209

Appendix: Debugging Case Study Solution — 213

Glossary   223

Index   229

Preface

This book is written as an introductory primer on the subject of structured program design. The reader is anticipated to be either a student learning about programming and using this material as a companion, or one who already knows the mechanics of a computer language and is now concerned about the design process. In either situation some background is assumed (previously or concurrently). Every effort has been made to present the material in a simple fashion without exotic notation or complex examples. We believe both administrative- and data processing-oriented individuals would profit from the material discussed which actually covers the spectrum from design theory through management process.

The material used in this discussion is commercial application oriented, however its value is not limited to this. Scientific applications typically require more exotic algorithms and less concern with input/output, but the program design needs are similar.

We are now some six years into the structured programming revolution, and as with most revolutions the ultimate winner-revolutionary or traditionalist has not emerged. Given past battles of this type we suspect a truce may be signed and a new battleground formed. Regardless of one's orientation, the argument against "good" programming and structure is as bad as an argument against motherhood. The more serious question is how does one accomplish it and what are the trade-offs (e.g., programming time, resulting efficiency, subsequent debugging and maintenance costs). Therefore, our role here is to introduce the key ideas of structured design in a readable manner and try to be objective about their merit. Each of the authors brings to the text a different background, but a remarkably similar philosophy about what is needed. The text remaining here had to pass a critical review by each coauthor.

Preface

Chapters 1 through 3 are designed to provide primarily history, philosophy and background needed for subsequent discussion. Chapters 4 through 6 describe, respectively, the subjects structured programming, structured design, and pseudo-coding. These chapters are quite mechanical and are geared to generally show *how to*. Keep in mind the goal here is to introduce the subject. Further reading and/or practice will be needed to become proficient in the various areas discussed. Chapter 7 is added to focus attention on the fact that while programming remains an art to some degree, there is considerable opportunity to facilitate structured design and subsequent debugging through good style habits.

Chapter 8 emphasizes the evolution of data base technology and its impact on the process. As one moves from the file-oriented traditional world to on-line data bases, certain new design considerations emerge. This chapter is kept purposefully abstract to maintain a broader perspective. It is hoped this approach will be adequate to facilitate a future exposure to some specific data base software package.

Chapter 9 describes how structured design and programming practices impact on the management planning, organizing and control process. Finally, chapter 10 summarizes the theme of this discussion and offers a prediction for the future.

Throughout the description of various concepts, it seemed necessary to demonstrate concrete examples by showing actual coding. We felt that three languages were viable candidates because of their widespread use or technical suitability. Our bias was that Programming Language I (PL/I) best fit the spirit of structured design and therefore it is used throughout as the illustrative language. Readers who know either FORTRAN or COBOL should have little trouble translating these examples into their "native" language. It is our belief that PL/I in spite of certain shortcomings is the best language with which to develop systems or programs which exemplify the tenets of structured design.

In various portions of the text we have made minor deviations from definitions or formats presented by others. We are aware of such changes and use them to make a point. It is important to realize that there is no god of structured design or czar of definitions. Rather, these are concepts, ideas and tolls produced by intelligent people to conceptualize a complex process. All methodology in this state of development is "soft" and should be allowed to evolve. We do not

Preface

intend any of the mechanical approaches to be the one best or only way to accomplish a particular goal. Common sense and intelligence of the users must dominate any set of so-called rules.

We hope readers will find after reading this text that they have a better perception of why structured programming became popular so quickly, what it is and, most importantly, what it is *not*. If this is accomplished, the effort is a success.

We are indebted to our students for their help and patience with the preparation of this material. Also, the design of the Force and Financial Planning System for the Air Staff, U.S. Air Force, provided a situation for many academic ideas to be tested in a real world environment. As with most ventures of this type, our families learned more about the subject than they cared to. We humbly accept responsibility for whatever errors remain.

Gary L. Richardson
Charles W. Butler
John D. Tomlinson

1
BACKGROUND

No discussion on structured program design could begin without some mention of the background and history of the structured programming revolution. Since the introduction of the "stored program," no other software concept has created such an impact on the data processing profession. Today, programmers, regardless of their native language or type of application, are divided on the issue of structured programming and the use of "GOTOs" in programs. Data processing managers are under increased pressures to produce software quickly and reliably, plus facilitate future maintenance of those programs.

Prior to December 1973, when *Datamation* published a series of articles proclaiming the "structured programming revolution," few people were routinely aware of its existence. Historically, structured programming origins can be traced back to Bohm and Jacopini's article in May of 1966 and E.W. Dijkstra's letter in March of 1968, both appearing in the Communications of ACM. Of these two historical "roots," it was Dijkstra's now famous "GOTO" letter that created the most controversial aspect of the structured programming revolution by his statement that GOTOs in programs were potentially dangerous and should be eliminated. Today almost everyone has heard about structured programming concepts. Unfortunately, no clear, precise definition of it is accepted universally.

Structured programming can mean a variety of program design concepts or techniques; it can be the specific coding structures of DOWHILE, IFTHENELSE or SEQUENCE; or the ideas and philosophy of top-down design/development; or hierarchy plus input-process-output (HIPO) charts, or Orr-Warnier diagrams, or pseudo-code; or the chief programmer team management process. But whatever the local "operative" definition of structured programming, the impact the revolution had will no doubt continue to affect software development for many, many years to come.

In the beginning (and not too distant past), many programming managers believed—and convinced their bosses—that programming was an esoteric art that could not be managed using normal management techniques. As a result, if software took longer to develop than planned, cost more than originally estimated, didn't meet user specifications, or suffered repeated failures upon implementation, the entire catastrophic production process was called "normal." Had we been developing anything else, instead of mysterious computer software, such performance by both labor and management would have created an outcry for solutions. Data processing, however, was the exception because its process of evolution was not well understood by either party:

> ... to talk about a software crisis was blasphemy. The turning point was the Conference on Software Engineering in Garmish, October 1968, a conference that created a sensation as there occurred the first open admission of the software crisis.[2]

Admitted or not, the software crisis has affected almost every software development project, both large and small. The large projects attracted the majority of attention. Two notable examples were IBM's OS/360 development nightmare, and the Air Force's Advanced Logistic System design which was ultimately scrapped after the expenditure of thousands of man-years of programming effort. Smaller projects suffered similar, less public fates recorded for the most part only in the memories of the scarred functional users or the "veteran" programmers. Frederick Brooks best describes the nemesis of the software crisis when he wrote:

> No scene from prehistory is quite so vivid as that of the mortal struggles of great beasts in the tar pits. The fiercer the struggle, the more entangling the tar, and no beast is so strong or so powerful but that he ultimately sinks. ... Programming over the past decade has been such a tar pit. ... Large and small, massive or wiry, team after team has become entangled in the tar. No one thing seems to cause the difficulty. ... Everyone seems to have been surprised by the stickiness of the problem, and it is hard to discern the nature of it.[1]

This analogy describes well the software development process as we often see it—confusing, complex, and often ill-fated.

Background

TRADITIONAL SOFTWARE DEVELOPMENT PROBLEMS

Plan, organize, control—these could be called the holy trinity of management. Planning and organizing are precise and straightforward concepts and are easily accomplished whether you are producing widgets or software. But controlling—how does one control a software development project? Frank Ingrassia describes the typical plight of the traditional software manager thus:

> One of the most common problems encountered in software management is the lack of visibility into progress being made by each individual on a project. Consider the "percent complete" estimates given by programmers. They are notoriously bad. . . . You may have the feeling of having sent the programmer into a dark, deep tunnel. . . . From time to time you might hear "I'm 90% through" or "I'm 98% through," but it's often a long time before the programmer emerges. In the meantime you find yourself spending a lot of time juggling assignments and commitments based on the assumption that the estimates are near the mark.[3]

Unfortunately, the software crisis affected both old and new software. The faulty design, typographical errors, and traditional poor/hasty coding techniques created during the original development process return time after time to plague subsequent maintenance efforts. From its creation to its demise, the typical production program spans at least ten programmer generations! Everyone knows of a program (or system) which was originally designed and coded for some long-gone mainframe that must now be emulated, translated, filtered, and converted through countless other mainframes/languages—and still is in active production. It's no wonder that programmers are often tempted toward complete redesign and reprogramming when faced with a particularly "messy" program to maintain. Faced with such a situation can there be any doubt that the software maintenance costs often account for 50% to 90% of the total life cycle costs of a program?

The software crisis is real. Managers and programmers both share the blame. Projects fail because managers don't manage their programmers, not because the programmers lack technical expertise. Programmers rush into coding, ignoring careful design and documentation because coding "looks busy" and designing "looks idle," or to insure that more time will be available for debugging. The "90% complete syndrome" is the accepted norm for too many software development projects. Why? Brooks again provides an illuminating comment:

> All programmers [and programming managers] are optimists. Perhaps this modern sorcery especially attracts those who believe in happy endings and fairy godmothers. Perhaps the hundreds of nitty frustrations drive away all but those who habitually focus on the end goal.[1]

EVOLUTION OF STRUCTURED PROGRAMMING CONCEPTS

No single person started the structured programming revolution; however, the first known recorded reference was in 1966 when Bohm and Jacopini provided the theoretical framework by showing it possible to write any program using only three logic structures: DOWHILE, IFTHENELSE, and SEQUENCE.[4] With these constructs it is possible to write programs that can be read from top to bottom without ever branching back. Using this approach the GOTO statement is not needed—Dijkstra's point in his famous "GOTO" letter.

The overlooked element of the great "GOTO controversy" is that when the three basic structures are used correctly, there isn't any occasion to use the "GOTO"! Around 1972, Harlan Mills of IBM further expanded the scope of structured programming by adding the requirement that a program have a single entry and a single exit point. Finally, the "classical" structured programming "rules" include a set of language-dependent conventions to dictate how to indent program statements which is designed to make the constructions more visible to the reader. Although (manual) indentation may in fact make the program initially harder to write (due primarily to old habits), the reading and understanding is greatly simplified which during program maintenance reading is especially critical.

Traditional programming projects have reported average coding rates as low as two to three statements per man-day; however, use of modern structured programming techniques has increased levels of productivity to near sixty statements per day. It is this need for improved productivity which has created the exploding demand for new design and coding techniques. The thrust of structured programming is to improve programmer productivity and decrease long-range software maintenance costs. Daniel McCracken, probably the best-known writer in the field says:

> It has been said that skilled programmers have pretty much been using

structured programming for years, anyway. This isn't really true. The discipline imposed by using only the three basic program structures and following indentation rules rigidly improves the performance of even the best programmers. Perhaps, even more important, it can greatly enhance the effectiveness of the rest of us, who are not geniuses and who sometimes program in rather sloppy ways. . . .[4]

Prior to December 1973, Dijkstra, Mills, Bohm, and Jacopini all were describing theoretical issues that didn't seem to apply to the real world of software development. Then the famous *New York Times* information bank project produced a complex system using structured programming and other contemporary tools. In 22 months, 11 man-years of effort produced 83,000 lines of high-level source code that was delivered on time and was found to be highly reliable. Suddenly, software managers and programmers were confronted with reports of vastly increased programmer productivity and greatly reduced coding error rates. The *New York Times* project involved structured programming, but it also included a new concept formulated by F.T. Baker—the chief programmer team, which essentially uses one "super programmer" to be the grand system architect and programmer. The success of this project indicated the potential for a new level of manageability in software development combining both technical and organizational innovations. For the data processing industry, this project became the industry's equivalent to the 1920s management theorists' experiments at the Hawthorne Western Electric plant.

IBM quickly capitalized on the techniques and organizational innovations of the *New York Times* project and applied them to the mission simulation system in the preparation and training for NASA's Skylab operations. In a two-year period, 400,000 lines of source code were produced using the structured programming and chief programmer team concepts. Remarkably, the software was delivered on schedule in spite of 1,200 formal changes in the requirements, manpower cuts, and reductions in the computer budget.

The December 1973 issue of *Datamation* formally started the structured programming revolution for the rest of the data processing profession. Soon, the basic structured programming constructs of Bohm and Jacopini, Mills' rule of single entry and single exit, Baker's chief programmer team, and Myers' top-down development concepts were supplemented by other "revolutionary" software tools. Numerous techniques and procedures have since emerged to assist in the design, development, and documentation of structured software. All of these techniques and procedures are called, or have been classified, as "structured programming." As a primer on structured program design, this text will describe the application of many of these techniques in the software design and development process.

The dictionary defines revolution as a complete change of any kind. The object of the structured programming revolution now addresses the entire set of processes, methodology, and procedures to create computerized systems. One central focus of this is the program development process and the nature of the product resulting from our programming effort. Structured programming concepts are dedicated to the production of "good" programs. What is a good program? To best answer this question, let us list six desirable qualities of a good program as follows:

1. *The program works*—the most important quality. Speed, resource requirements, number of source lines, or any other measurement variable are all meaningless for a program that does *not* work. In the book *The Elements of Programming Style* by Kernighan and Plauger, three rules are cited as follows:
   a. Make it right before you make it fast.
   b. Make it failsafe before you make it faster.
   c. Make it clear before you make it faster.
2. *Lower testing costs*—30% to 50% of total project time is devoted to testing and debugging computer programs. An adequate level of testing is required to avoid catastrophic errors during production.
3. *Lower maintenance costs*—data processing organizations spend 50% to 90% of their annual software costs on the maintenance of existing systems. The high cost of maintenance limits the development of new software.
4. *Ease of modification*—no matter how carefully planned and designed a program is, there will always be changes and modifications. Every segment, module, or procedure of a program should be designed with an eye toward its eventual revision or change.
5. *Uncomplicated design*—programs need not be complex to work properly. In fact, the most logical way, and in some cases the only way, to make a program easy to test, maintain, and modify is to keep it simple. *Always*

design a program so that someone else can maintain it.
6. *Efficiency*—normally 50% of the execution time is used by approximately 3% of a program's instruction set. The following methodology should be used to achieve higher performance:
   a. Write the program's logic in a straightforward manner, emphasizing simplicity and reliability.
   b. After the program is working, rewrite and optimize the time-consuming code.

Using this method programmers will invest their time optimizing only the code that saves an appreciable amount of time. Keep in mind that the optimization process *may* cloud the design structures.

STRUCTURED PROGRAMMING TECHNIQUES

Although there is controversy as to whether the chief programmer team can be implemented in some development environments due to rigid personnel policies, we have chosen to keep it in the list of contemporary program design concepts. Basically, this text recognizes seven essential ingredients to structured program design. These are described briefly here and will be the specific subject of subsequent chapters:

1. *Top-down design.* Top-down design is the process of decomposing a single complex function into hierarchical levels of simpler functions. The levels of the hierarchy correspond to the control levels of the tasks performed by the system components. Each level of the software design shall be logically complete in itself. The top level contains the highest level of control logic and decisions within the software design. Each sublevel is a self-contained component whose operation is subordinate to the next higher level. In this fashion, the overall design is decomposed into more and more specific logical segments.

2. *Top-down documentation.* Top-down documentation mirrors the top-down design and is delivered in increments as the system is developed. It records top-down design decisions, insures that those decisions are

made prior to coding, and serves as the review/approval document to authorize coding. It includes functional descriptions, input/output specifications, graphical representations, operations manuals, reports, listings, and other technical documents. Some new techniques of documentation currently emerging are hierarchy plus input-process-output (**HIPO**) charts, and pseudo-coding.

3. *Top-down implementation.* This involves the coding, verification, and implementation of higher system logic levels prior to coding of any subordinate modules. Lower-level modules, needed for an interface, are coded as dummy code (stubs) which need not perform any meaningful computations. The important point is that program modules at each level are fully integrated and verified with their predecessors before coding begins on the next lower level.

4. *Structured programming constructs.* The techniques of coding programs using only the control logic primitives: SEQUENCE, IFTHENELSE, and DOWHILE. This discipline allows *no* other coding constructs, except the two additional structured programming constructs.

5. *Structured walk-throughs.* Internal technical examinations of the design, implementation, and documentation of a program or system are made by the design project staff to provide positive feedback to the programmer/designer. These walk-throughs are scheduled by the programmer/designer and attended by his peers. This epitomizes Weinberg's "egoless" programming concept.[5]

6. *Pseudo-coding.* An English-like language used to describe the control structure and general organization of a computer program. It is designed to be easy to read and comprehend by others. In addition, it is structured in the sense that it utilizes a semi-formal grammar and syntax which incorporates structured coding constructs.

7. *Chief programmer team.* A method of staffing a software development project, whereby a single "super programmer" is assisted by a number of other

programmers and data processing technicians. The chief programmer is the primary architect for the project's design and implementation.

It is important for the reader to realize that these techniques are in a floating state both in terms of the overall concept and its general acceptability to the practitioner. A recent professional association program announcement epitomizes this state of fluctuation. The announcement went something like this: Mr. X will discuss his implementation experience with top-down structured programming; four years earlier Mr. X had called structured programming an unworkable fad and ran the most archaic data processing operation one might wish to see. Thus, our approach here is to discuss the various techniques and ideas which constitute contemporary structured program design without regard to too many "war" stories. We have found the concept workable but as with all change it is not simple to implement into a traditional environment.

Review Questions

1.1. Outline the key events in the history of structured programming.

1.2. Is the structured programming revolution primarily a programming discipline or a management technique?

1.3. What are some of the key reasons for the rapid acceptance of structured programming?

1.4. What are the seven essential ingredients of structured program design?

1.5. Is there a software crisis today? Why?

1.6. Why are the *New York Times* and Skylab projects so important to the acceptance of structured design concepts?

1.7. When is execution efficiency addressed in the structured design cycle? Do you agree with this (try to address both pro and con issues)?

Notes and References

1. Brooks, Frederick. *Mythical Man Month: Essays on Software Engineering* (Reading: Addison-Wesley Publishing Co., 1975).
2. Dijkstra, Edsger W. "The Humble Programmer," *Communications of the Association for Computing Machinery* 15 (October 1972): 859–866.

3. Ingrassia, Frank S. "Combating the '90% Complete' Syndrome," *Datamation* 24 (January 1978): 171–177.
4. McCracken, Daniel D. "Revolution in Programming: An overview," *Datamation* 19 (December 1973): 50-51.
5. Weinberg, Gerald M. *The Psychology of Computer Programming* (New York: Van Nostrand Reinhold Co., 1971).

2
EVOLUTION OF PROGRAM DESIGN METHODOLOGIES

An examination of the current literature provides the reader with a proliferation of material pertaining to program design methodologies. Of course, knowing the current state of the technology is not the only requirement for successful implementation. For the sake of continuity, it seems appropriate to describe how today's state of the art has evolved from its infancy. By examining the design evolution, the factors which have influenced change can be identified and their impact analyzed.

In this chapter, a historical perspective of system design is presented. The evolution is traced from the 1950s when computer-based systems began to emerge through today's contemporary design methodologies. For selective design techniques, a discussion is provided to orient the reader with each separate approach to problem analysis.

In recent years (since 1973), there has been increased interest in the introduction of new approaches to design problems. The exact cause of this influx of new design methodologies was briefly discussed in chapter 1 and are closely linked to the rapidly expanding cost of software and marked improvements in problem understanding. The scope and pace of these technological changes are unprecedented in history. Socially, we are more educated and our tastes for consumer products and services are more diverse than ever. In responding to a more sophisticated public, business has found that many of the past methods of operation are insufficient when dealing with the market framework as it exists today. Given the risk of decision making in a rapidly changing market, the business firm is faced with the need for higher quality and more timely information. In many ways this can be defined as an information revolution and software development techniques are lacking to support the increasing needs and evolving

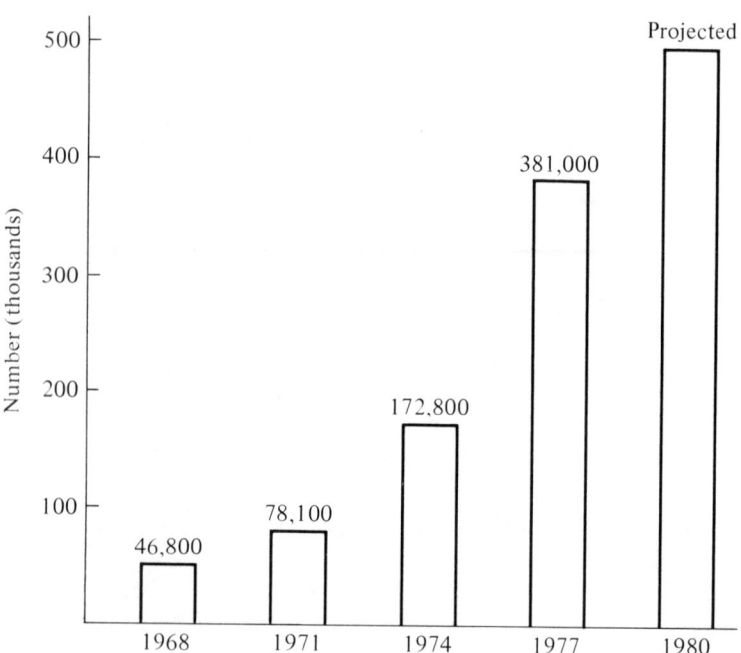

Figure 2.1: Trends in computer units installed

computer capabilities. At all levels, management's informational needs have increased and the information processing capability of the computer is the tool most have turned to.

The revolution of computer technology and the complexity of the business environment go hand in hand. The computer is a tool which has benefited greatly by technology advancement and, at the same time, has contributed to advances in all fields. Any examination of computer technological advances requires a look at both hardware and software elements owing to the interrelation between these components.

Computer hardware consists of all the physical components that make up a functioning computer system. Since completion of the first effective computer, ENIAC (Electronic Numerical Integrator and Calculator) in 1946, hardware developments have generally resulted in smaller components, greater speed, lower costs, greater storage capacity, and improved reliability. The ENIAC weighed a tidy 30 tons, took up 1,500 square feet of floor space, and contained 19,000

Evolution of Program Design Methodologies

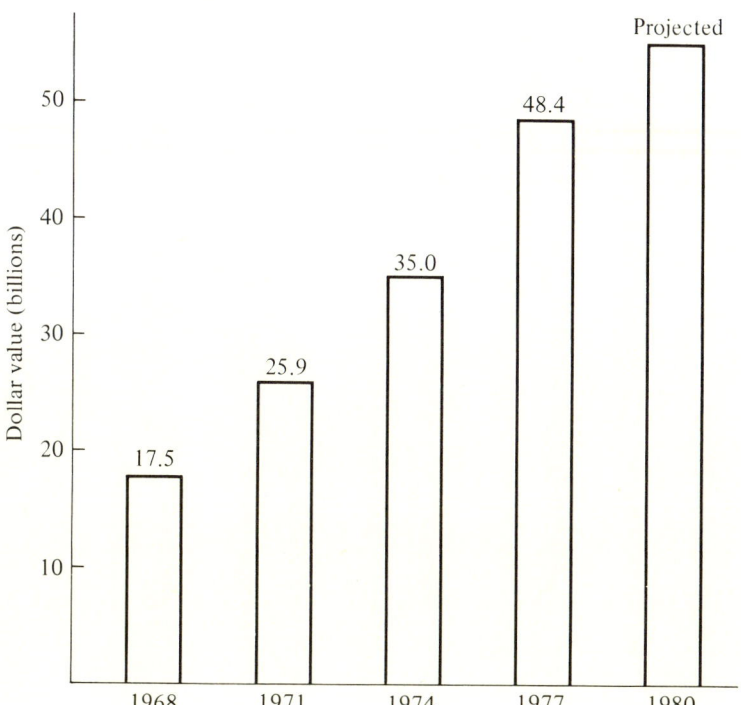

Figure 2.2: Dollar value of computers installed

vacuum tubes. Today, greater computing capacity is achieved with a "computer-on-a-chip" technology. Coinciding with technological innovation, computer utilization has grown exponentially. Figure 2.1 provides a summary of the number of physical computer units installed during the period from 1968 to 1977, and Figure 2.2 illustrates the dollar value of this growth. The number of computers installed between 1968 and 1977 grew over 700%, while their dollar value increased by 175%. This implies that average unit cost has declined and will continue to do so in the foreseeable future.

Computer software consists of all programs associated with the use of computer hardware. When compared to the tremendous hardware advances, it becomes apparent that the evolution of software has lagged behind hardware developments. In fact, in the area of system design methodology, it is a generally accepted fact that design techniques lag the hardware evolution by one full generation.

Figure 2.3: Software as a percentage of total computing cost

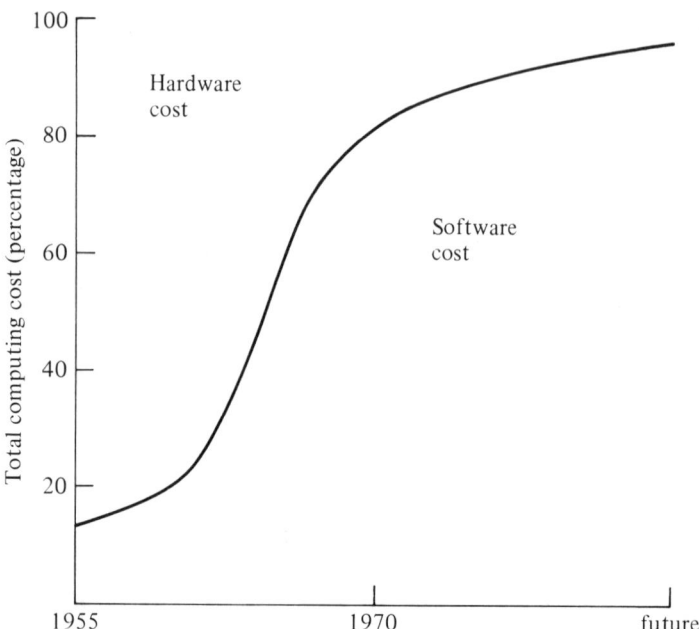

Figure 2.3 illustrates the hardware-to-software cost ratio trend since 1955. From this it is easy to see why management's emphasis has shifted toward techniques which provide efficient program preparation and improved control. What then have been the major factors determining the need for improved design techniques?

First, hardware changes impact software systems. In the past, a change to, or update of, an existing hardware component might have resulted in the need to completely prepare new programs to replace existing operational systems. Therefore, design techniques which encompass the entire scope of a system serve to ease transition problems.

Second, firms operate in an increasingly complex environment. Integrated systems are needed to produce information covering the firm's entire operating sphere. In the past, systems were designed with limited scope, such as a payroll system or an inventory system. Today each of these is merely a subsystem, a single element of another larger system. Data required by one subsystem is also needed by numerous other entities throughout the total system. Thus we find an increasingly integrated and hierarchical approach to sys.ems design which is far different from the design of a file-oriented payroll system in the early sixties. Today we are concerned with data base design and integration more than record layout and low-level logic. In such an environment, good design techniques must facilitate systems integration by examining the relationships between individual components.

Finally, languages used in the implementation process have modified the nature of the programming effort. Language processors have evolved from low-level machine language to high-level procedure-oriented languages; consequently, less emphasis is being placed upon machine efficiency and more concern is placed upon reduction of program preparation costs and time. In the modern environment, design techniques must assist in reducing the initial development, as well as the subsequent maintenance costs. Emphasis has now shifted away from highly mechanical issues of the host machine itself to more complex functional relationships involving the user process.

EARLY DESIGN TECHNIQUES (BEFORE 1960)

Let us now take a look at some of the early design techniques. An extensive study conducted by Cougar traced the evolution of business system analysis techniques, and it is used as a basis for much of the following discussion.[1] This chapter will describe how various design techniques have viewed the software development process historically. The following seven steps of system development are used:

1. Documentation of the existing system
2. Analysis of the system to establish new requirements
3. Design of a computerized system to meet requirements
4. Programming and system development
5. Implementation
6. Operation
7. Maintenance and modification

Software design is, then, concerned with all of these steps at some time.

Starting with the 1950s, design techniques for analysis of computer-based systems began to emerge from earlier industrial engineering tools. The concept of system analysis as applied to computer technology at this phase of evolution was primarily concerned with

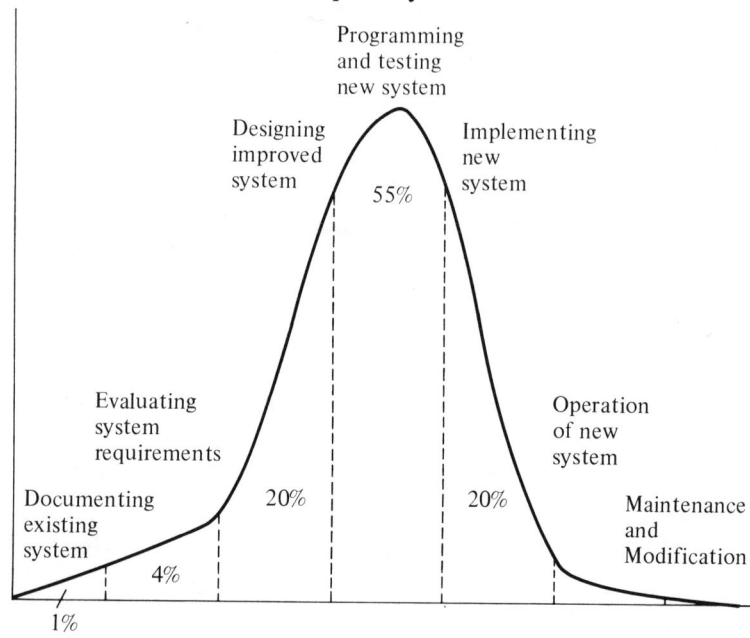

Figure 2.4: Relationship of design stages during early computer systems

documentation of existing manual procedures (e.g., payroll, accounting, inventory, etc.). Analytical tools in this period concentrated on the flow of information through the organization, especially as it applied to operational-level systems. Generally, these systems were designed and implemented with little regard to interdependence with other systems. This lack of concern for integration minimized the front-end cost of design, but obviously complicated future changes. As firms identified functions which were suitable for computerization, increased efforts were made to create system after system with no apparent concern for an overall development structure. With increased utilization of machine time, businesses were able to justify the expenditure for new hardware. It must be remembered that during this time period machine cost was the major portion of the development budget (see Figure 2.3). As a consequence, little attention was paid to efficient software design and, as shown in Figure 2.4, a major portion of resources consumed was concentrated in the middle stages of the process (e.g., programming and testing).

During the 1950s three techniques emerged as classical analysis tools. These are:
1. Flowcharting
2. Information process charts
3. MAP system charting techniques

Flowcharting

Flowcharting techniques had existed for several years, but as applied to data processing they evolved into a graphical tool which provided a means of recording, analyzing, and communicating problem information. It was, for the analyst, a schematic for recording the flow of logic in an operating procedure from its originating source, through various processing operations, to the final output. At the system level, the flowchart provided a broad overview of the processing operations to be accomplished, while at the program level it presented a detailed graphical schematic of logic steps for the computer to perform in order to generate the desired output. Undoubtedly, this is the most used design tool both historically and today, yet it does little to enforce structure on a design.

Information Process Charts

An information process chart (IPC) is a combination of flowcharts and block diagrams. IPC conceptually contains verbs and comments which were carefully defined to insure understanding among all users. Such standardized illustrations were useful for computer analysis, and IPC was the first technique to recognize the need for such formal construction.

MAP

The MAP system charting technique ws developed by NCR and it permitted an overview of information flow. An important element of MAP was a "transaction break" which showed interrelationships between files. Unfortunately, MAP procedures provided little standardization for constructing the design structure and, as a result, it provided little foundation for subsequent analysis steps.

DESIGN METHODOLOGIES OF THE 1960s

In the 1960s computer hardware technology advanced rapidly. With enhanced hardware performance, systems analysis techniques began to take on new emphasis. The software development role began to take on an expanded scope in meeting growing managerial information needs. Design techniques began to exhibit an increased orientation toward an overall system definition.

Information Algebra

In 1959 the Language Structure Group of the CODASYL Development Committee set about to develop a structure for a machine-independent problem-defining language. This product, Information Algebra, was published in 1962. Basically, its role was to extend the concept of stating relationships among data and for all aspects of data processing. It utilized modern algebra and point set theory as a conceptual foundation for its methodology and had the following assumptions:

1. Information deals with objects and events relative to the task at hand.
2. The objects in the system, defined as entities, are represented in the system by data.
3. Outputs from the information system can be extracted by definable processing logic.
4. The significance of a particular entity is derived from a data value which describes its associated attribute of property.

Using Information Algebra, the analyst specified the relevant sets of data and then defined relationships and rules of association by which the data sets were to be manipulated. Through these relationships new sets of data are developed, including the required output. Within this methodology, system optimization was sought by providing a means to define a system. Although the method itself never had a great impact on actual techniques, it showed that design researchers were beginning to look for methods to produce designs using shorthand notation and that the need for formal understanding of logic processes was being recognized.

Study Organization Plan

Study Organization Plan (SOP) was developed in 1961 by IBM as a method of gathering data and using this data to analyze the information needs of the entire organization. Consistent with the expanding concern for scope and integration, SOP was organizational oriented. When using this technique, the firm organized data into three distinct elements: general, structural, and operational. General data contained a history of the firm, an industry profile, organizational goals and objectives, major policies and practices, and relevant governmental restrictions. In contrast to general information, structural information presented a concise picture of the business. It included detailed data pertaining to the firm's products, markets, materials, suppliers, finance, personnel, facilities, and inventories. Finally, the operational element provides data describing the operating activities of the business such as flow diagrams and a distribution of the total resources within the operating activities. As can be seen from the brief description of this technique, the emphasis developed here was of an information orientation rather than logic activities. The complexity and risk associated with the business environment required the designer to become more knowledgeable of all aspects of the firm's operation, as well as its economic environment.

Accurately Designed System

Accurately Designed System (ADS) was developed by NCR in the late sixties. Within this design framework there are five formal steps. First, outputs were defined. Second, inputs necessary to create the required output were formally identified. The third step was an identification of computations needed to convert given input into required output. Elements of computation definition included logic restrictions on the computation process, significant interrelationships of the computation process, and a definition of various information sources used in each computation. Another step of ADS was the history definition which identified data to be retained beyond a given processing step for a subsequent step. Finally, ADS required a formal logic definition in the form of a decision table.

In addition to the formal design steps described above, there were other important operational procedures within ADS to establish information linkages. Each data element was identified and assigned a reference name. Subsequently, each time the data element was used in the system it was linked back to the previous reference or source. Through this procedure a chain was created for each element detailing its flow from input to output. In many ways this technique was the predecessor of a number of the top-down design tools which emerged in the early seventies.

COMPUTER-ASSISTED DESIGN TECHNIQUES

It can be concluded from the design philosophies of the sixties that development scope was rapidly shifting from implementation of small localized systems toward analysis of whole organizations and the need for integrated relationships. Stimulated by this, the scope and sophistication of resulting systems required improvement in development techniques. An immediate result of this growing complexity was the employment of the computer as an integral element in system development. Consequently, logic processors were created to assist the system designer. Three techniques will be discussed below to exemplify this thrust. The reader should recognize that other efforts are currently underway.

Decision Table Processors

The earliest attempts were in the area of decision table analysis. In 1965 the Special Interest Group for Programming Languages (SIG-PLAN) of the Association for Computing Machinery (ACM) developed DETAB/65. This processor accepted decision tables as input and converted them into COBOL source code. Once again we note the emphasis on total system development, but now there is a concern for coding, development, and maintenance of the software. In some ways this is a meta-language approach to design.

Automated ADS

Decision table processors fell short of providing an integral analysis technique for information systems. However, automation of ADS was an initial approach designed to achieve system integration by computerizing the procedures of ADS. The output was system documentation which was useful for lower level design. Such factors as speed, accuracy, and timeliness were identified. In addition, the importance of system maintenance and future modification was being recognized as a part of the original design process. Since the ADS definitions were in machine-readable form, revisions or modifications could be easily made and this benefit upon the process was being increasingly recognized.

Time Automated Grid

In 1966 the idea of computer-assisted analysis techniques was developed further with the implementation of the Time Automated

Grid (TAG) system by IBM. There were several programs within this system which assisted the analyst in developing system specifications. By working backward from the required outputs, necessary inputs were defined. As they were developed, file formats and definitions could be produced. The output of the TAG system included a series of reports summarizing the system definition. The most important output was the time grid analysis which traced the appearance of each data element, by time, through the entire system.

Development of computer-assisted analysis techniques were significant in the evolution of system design methodologies. However, it became apparent that the complexity of the business environment did not lend itself to complete automation and associated standardization of the analysis process. As the seventies approached, many computerized systems had been in operation for a decade or more and the need for substantial modification grew. By the mid-seventies there was a concrete realization that something must be done to alleviate the ballooning cost of software development.

CONTEMPORARY DESIGN TECHNIQUES

All of the current design methodologies appear to have a common ingredient, that is, each attempts to solve problems by providing an architecture from which the analyst can dissect and decompose even the most complex situations into one of manageable size. A structure for design today includes procedures, standards, and other supplementary tools which aid the analyst in interrogating all phases of the design effort. In particular, great emphasis is being placed upon earlier phases of the design effort (documentation of existing systems and analysis of the existing system to establish requirements for improvements) and later stages (maintenance and modification). The reason for this can be seen in Figure 2.5. With the complex and dynamic nature of business, increased time is being spent developing systems. In addition, business dynamics require continuous refinement of an existing system so that it meets the needs of its changing operational environment. Let us examine selected contemporary design concepts of Warnier, Jackson, and Ledgard. Each of these individuals produced concepts which set the stage for the structured programming revolution that began in the United States around 1973.

Figure 2.5: Relationship of design stages for modern computer systems

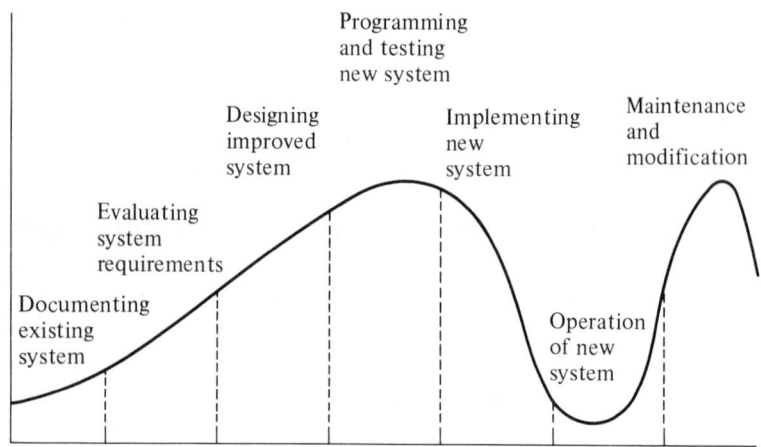

Logical Construction of Programs

Logical Construction of Programs (LCP) was originated in the early seventies by Jean Dominique Warnier in France. Figure 2.6 schematically illustrates this design approach. When applying LCP the user identifies all input data and organizes it into a hierarchical relationship. Each logical entity is identified. The number of times it occurs and conditional requirements for its execution are specified. These design procedures are repeated until the desired output structure is achieved. Having developed the required input and output data structures, a macro view of the design is produced showing input/output, condition operations, important calculations, and subprogram interfaces to be performed by the procedure.

Next, in a flowchart-like manner, the logical sequence of the macro-view design is established. During logic development, an important design tool, the Warnier Diagram, is utilized. This diagram is constructed with a series of brackets used with a small number of symbols to decompose a problem. By using this schematic, sets of actions are defined and each is noted with a beginning and an ending indicator. Important characteristics and interrelationships between actions are developed. Each element of the logical structure is expanded to achieve detailed development. For example, if a series of numbers are to be aggregated and a total printed, then a Warnier

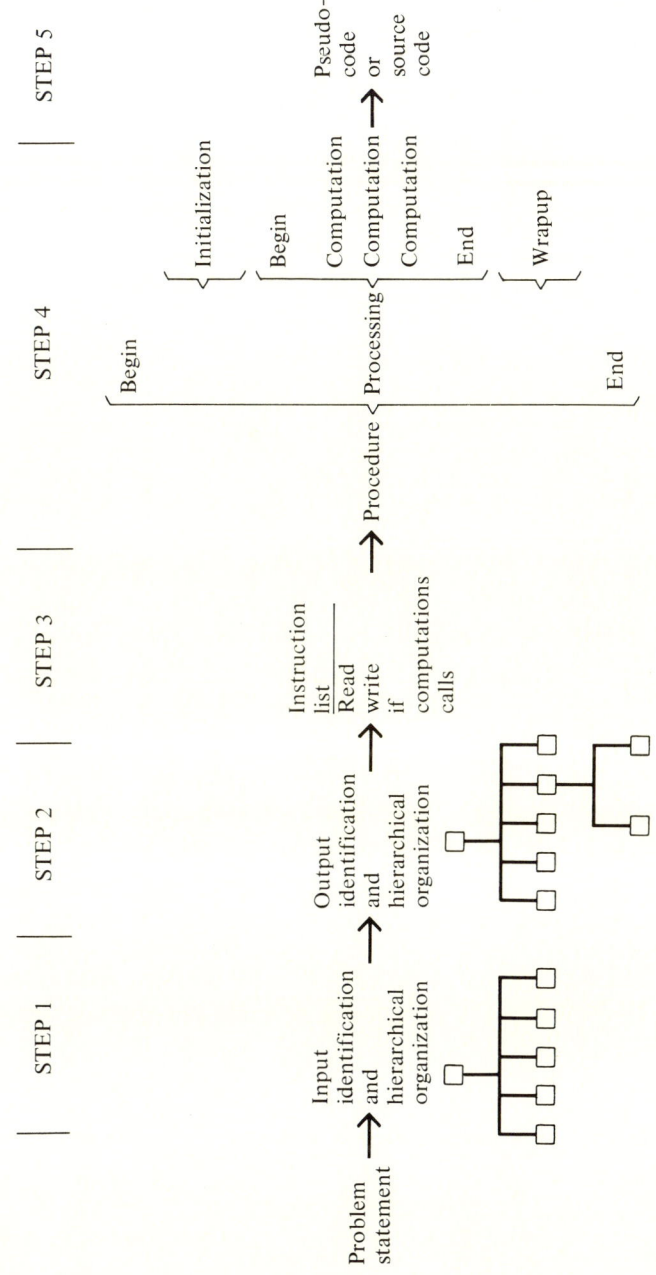

Figure 2.6: Logical construction of programs (Warnier approach)

A Primer on Structured Program Design

Figure 2.7: Warnier diagram for the summation of a series of numbers

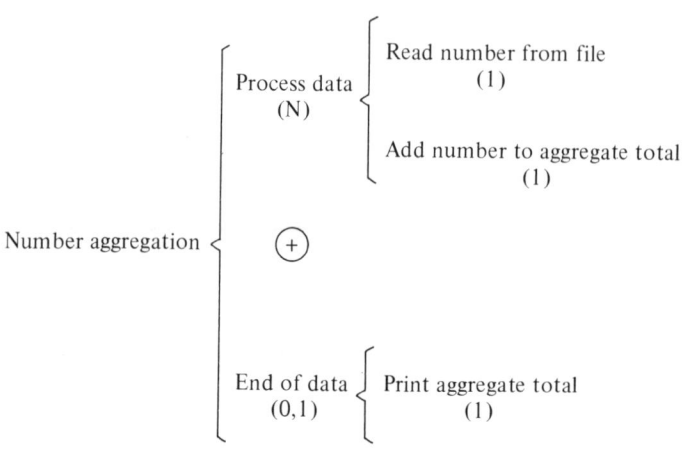

Diagram for this is illustrated in Figure 2.7. A detailed explanation of Figure 2.7 follows:

PROCESS DATA (N)	*means*	The logical entity "process data" is executed N times at this point.
END OF DATA (0,1)	*means*	The logical entity "end of data" is executed 0 or 1 time depending if there is an end-of-file condition.
+	*means*	The logical entities "process data" and "end of data" are mutually exclusive and only one of them will be executed.
READ NUMBER FROM FILE (1)	*means*	The logical entity "read number from file" is executed 1 time at this point.
ADD NUMBER TO AGGREGATE TOTAL (1)	*means*	The logical entity "add number to aggregate total" is executed 1 time at this point.

PRINT AGGREGATE   *means*   The logical entity "print
    TOTAL   aggregate total" is executed
     (1)   1 time at this point.

In a Warnier Diagram, the logical sequence is depicted by a left-to-right, top-to-bottom movement. This method appears to be well suited for simple, as well as complex, problem analysis since one or several modules can be developed easily.

Jackson Methodology

A method similar in nature to LCP is the Jackson Methodology developed in England by Michael Jackson. This design approach views a program as the means by which input is converted into required output, and, most importantly, it views data structures as driving forces for successful design. Figure 2.8 outlines three steps in the design process. First, the problem environment is considered and an understanding recorded by defining structures for data to be processed. Second, a program output structure is formed based on the defined data structures. Third, the task to be performed is defined in terms of elementary operations available. Each of those operations are allocated to suitable components of the program structure. Most importantly, program structure is associated with some component of the data structure. In many instances there is not a direct transformation of an input element to an output element. To resolve these "structure clashes," intermediate elements and files are often required;

Figure 2.8: The Jackson methodology

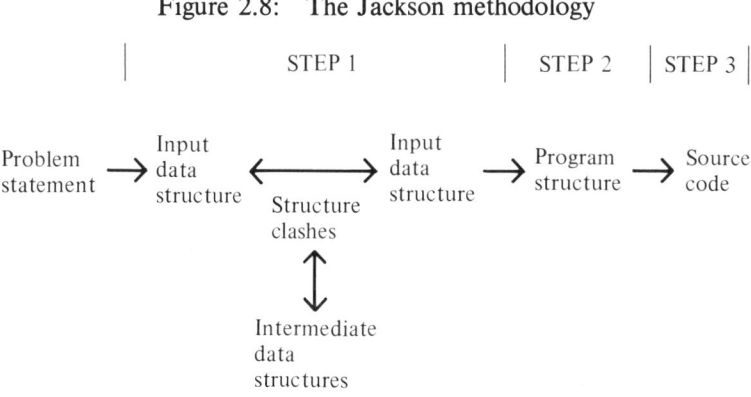

however, if the design can preclude the use of intermediate files, they should be avoided.

As with many contemporary design techniques, hierarchical structuring is an important element. When forming a program structure, large and complex problems can be reflected in the breadth and number of levels in the hierarchy. This hierarchy consists of both elementary and composite logic components. Elementary components are single instructions which do not require further development. They are, in fact, directly transferable to source code. The latter, composite components, consists of coding necessary to bind together various subparts. According to Jackson, there are the following three types of composite logic components:

1. Sequence—two or more parts occurring in order
2. Iteration—one part occurring one or more times
3. Selection—two or more parts of which one and only one occurs once

It should be apparent to the reader that there is a "commonality" between the Jackson and Warnier methodologies. The reason for this

Figure 2.9: Meta-stepwise refinement

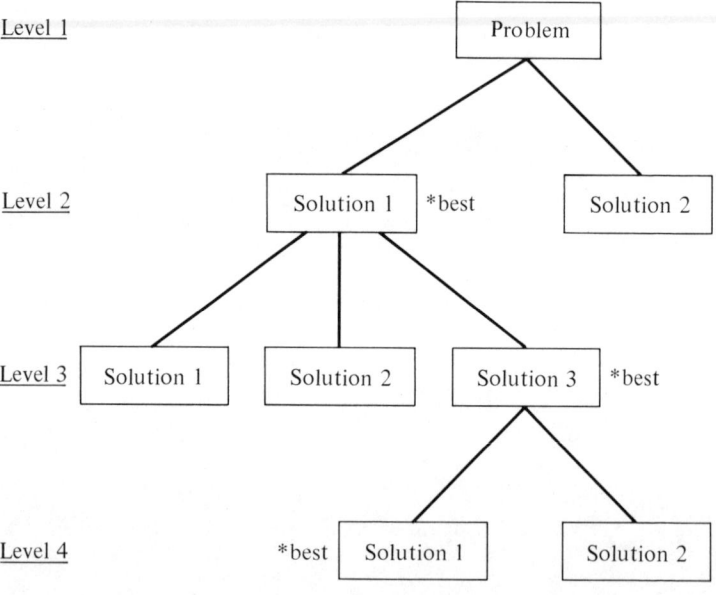

is inherent in their design philosophies. Each believes that the identification of the data structure is vital, and the structure of input/output requirements can be used to derive the logic structure of the application program. For well-organized problems this assumption will generally hold true.

Meta Stepwise Refinement

Meta Stepwise Refinement (MSR) differs from Jackson and Warnier methodologies because it emphasizes the process by which the designer generates the solution to a problem statement. Authored by Henry Ledgard, MSR is a composite of top-down notions, stepwise refinement, and level structuring (discussion of these topics is deferred until later chapters). The basic premise of MSR allows for the solution of complex problems by assuming a simple high-level solution, then gradually building in more detail until a complete solution is derived (see Figure 2.9). As detail is developed, several refinements or alternatives are generated for each level of detail. Only the best of these is utilized and it serves as a source for greater refinement.

FUTURE TRENDS

One conclusion to be drawn from the discussion of current design techniques is that their emphasis is on proper definition and analysis of data and logic prior to program coding. As a result, there is a natural evaluation of problem solutions including input/output specifications and logic development. Inherently, documentation is provided to aid in future maintenance needs, and by requiring development at the lowest level of detail initial implementation of the system is normally facilitated. Also it should be noted that none of these techniques requires explicit programming ability until the final stages. Thus, within the scope of system design, emphasis is shifted toward the earlier conceptual phases.

Since design methodology has expanded in scope to include all stages of the design cycle, what then does the future have in store for the designer? First, allow us to assume that no single technique now exists which is applicable to every design problem, so continued refinement of current techniques will occur in the search for utopia. However, enough research has transpired to indicate the direction of the next step in the evolution of system design techniques. As there was development of automated processors to assist earlier system analysis, there will be further research which develops automatic processors for each phase of system development. The objective of high-level processors will be to produce complete, integrated systems. In some predefined form, management will structure a problem and present it to a processor. The processor will then analyze the problem statement and provide management with the following automated outputs:

1. Input specifications
2. Output specifications
3. File (data base) structures
4. Logical constructs
5. Pseudo-code or source code
6. Hardware requirements
7. Potential problem areas

To the novice, high-level processors might appear technically infeasible, and it is not our premise here that system design will ever become totally automated. However, we do feel the total scope of the design cycle will be closely scrutinized and the computer will be utilized when possible to increase the speed, quality, and timeliness of system generation.

SUMMARY

As current design methodologies are changed and future ones are introduced, designers share a common goal—the development of procedures for solving large complex problems in an orderly fashion. The contemporary methodologies described here, especially those of Warnier, Jackson, and Ledgard, place major emphasis on hierarchical organization of both data and programs. The subject, structured program design, presented in the following chapters shares a common design philosophy with the notions of these individuals. It too places major emphasis upon decomposing complex problems in a structured manner. Structured program design is based on concepts published by Yourdon, Mills, Myers, Constantine, and Stevens, among others. The techniques discussed in the following chapters all have the goal of producing systems which exhibit the following characteristics:

1. High degree of structure in the design stages

2. Controlled branching at pseudo-code and source code levels
3. Improved level of understanding by users, managers and technical designers
4. Scrutable programs, i.e., programs consisting of code that can be understood by humans with reasonable intelligence
5. Enhanced ability to be maintained and modified
6. Improved degree of accuracy and conciseness of program design

Chapters 3 through 7 present pragmatic approaches for structured system and program design. Throughout these discussions, the techniques are developed without regard to any one individual's specific terms or philosophy. Rather, our goal is to present a coherent explanation of relevant theories and workable approaches. It is not a single gospel according to Warnier, Myers, Mills, or any other contemporary writer, but certainly the message is consistent. A serious reader should seek out specific reference material accompanying each chapter to delve further into individual philosophies.

Review Questions

2.1 What operation functions were computerized during the 1950s? What were the underlying reasons for computerizing these functions?

2.2 What type of computer expertise was required during the initial computer years? Does today's designer utilize similar computer knowledge?

2.3. Flowcharting is still a popular tool. Why has it survived the design revolution?

2.4. Are there any common characteristics between Study Organization Plan (SOP) and Accurately Designed System (ADS)? Between ADS and the Warnier methodology?

2.5 List five steps performed when using ADS.

2.6 When performing maintenance on a system, what factors would a designer have to consider?

2.7 Using Warnier diagrams, develop logic to find the maximum and minimum numbers from a series of numbers.

2.8. What unusual characteristic does Meta Stepwise Refinement (MSR) exhibit? What is its implication?

2.9. What were the design objectives for systems during the 1960s?

2.10. How is the design effort historically tied to the profit motive of the firm? Discuss.

Notes and References

1. Cougar, J. Daniel. "Evolution of Business System Analysis Techniques." *Computing Surveys,* September 1973, pp. 167-198.
2. Jackson, MA. *Principles of Program Design.* New York: Academic Press, 1975.
3. Ledgard, H.F. "The Case for Structured Programming." *Bit,* vol. 13, 1973, pp. 45-57.
4. Orr, Kenneth T. *Structured Systems Design.* Kansas City: Langston-Kitch, 1978.
5. Peters, Lawrence J. and Tripp, Leonard L. "Comparing Software Design Methodologies." *Datamation,* November 1977, pp. 89-94.
6. "Structure Systems Development." *Computer Decisions,* August 1977, pp. 26-29.
7. Warnier, J.D. *Logical Construction of Programs.* Leiden: Steinfert Kroese, 1974.

3
TOP-DOWN DESIGN

Because the development of software systems is increasingly viewed as a management problem, it has become a common objective to create a design philosophy which can provide an economic solution to the accelerating cost of software design and maintenance. One of the ideas surfacing from this research is that computer programs written with a high degree of structure facilitate testing, maintenance, and modification, and as a result, the concepts of top-down design and structured programming have gained rapid and broad acceptance throughout the data processing industry.

The concept of structured programming has evolved from philosophies such as modular and top-down program design. Traditionally, a set of interrelated logical units were developed separately and linked together to form a complete program. In top-down design, logical segments are defined first at high control levels and then successively lower levels of logic are detailed. However, unlike traditional programming, the top-down methodology implies that program logic is constructed with standardized, controlled branching which can be essentially read from top to bottom. A comparison of these approaches reveals that the latter is better suited to the task of creating error-free programs, since significant high-level code and important interfaces are tested with each implementation of lower level program segments.

This chapter and the one following are written, respectively, as macro and micro views of the design process. This chapter will use the "black box" approach to discuss how broad logic segments should be generally conceptualized and organized into a working system. Chapter 4 will look inside the logic boxes (segments) and discuss how their coding structures should be created. Both of these discussions will be kept at a high level of abstraction to simplify the concepts.

Collectively, these two subject areas form a major part of the theoretical foundation for structured program design.

TRADITIONAL APPROACH

All program design typically starts from one common point—the designation of functional requirements and specifications for a given set of outputs. After this step, the traditional approach has a distinctive set of events. First, all functional modules in the program are identified, defined, and coded. Each unit is then tested and debugged and, upon completion, integrated into one system. Past history has shown that, when utilizing this design methodology, definition and coding are often performed as unique operations. Also, module interrelationships may have been simultaneously defined by more than one person and are often inconsistent. Experience has proven that the integration of modules for testing results in problems which were overlooked in the design step. Subsequently, testing to integrate the modules uncovers clerical errors in coding and logic errors in design, plus other errors which may cause major revamping of the design. Thus, the traditional development process resembles the diagram illustrated in Figure 3.1.

TOP-DOWN APPROACH

In contrast with the traditional approach, top-down design starts by

Figure 3.1: Practical relationship of the design steps using traditional design methodologies

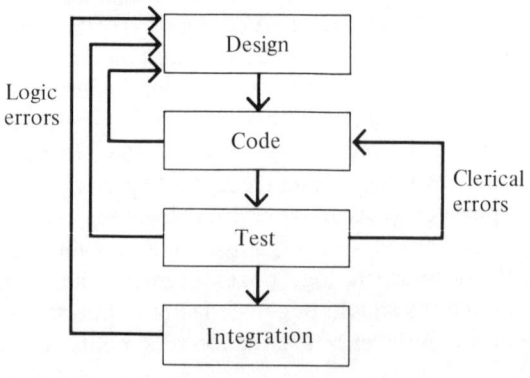

Top-Down Design

coding high-level logic segments first, then progressively expands this code until the lower levels are written and tested. As a result, testing and integration are minimized during system development (see Figure 3.2). Top-down implementation proceeds in a way that allows for continual integration of unique segments as they are coded and logical linkages are created prior to the individual parts being programmed. This means the critical top level logic is tested during each execution.

Figure 3.2: Comparison of testing and integration resources

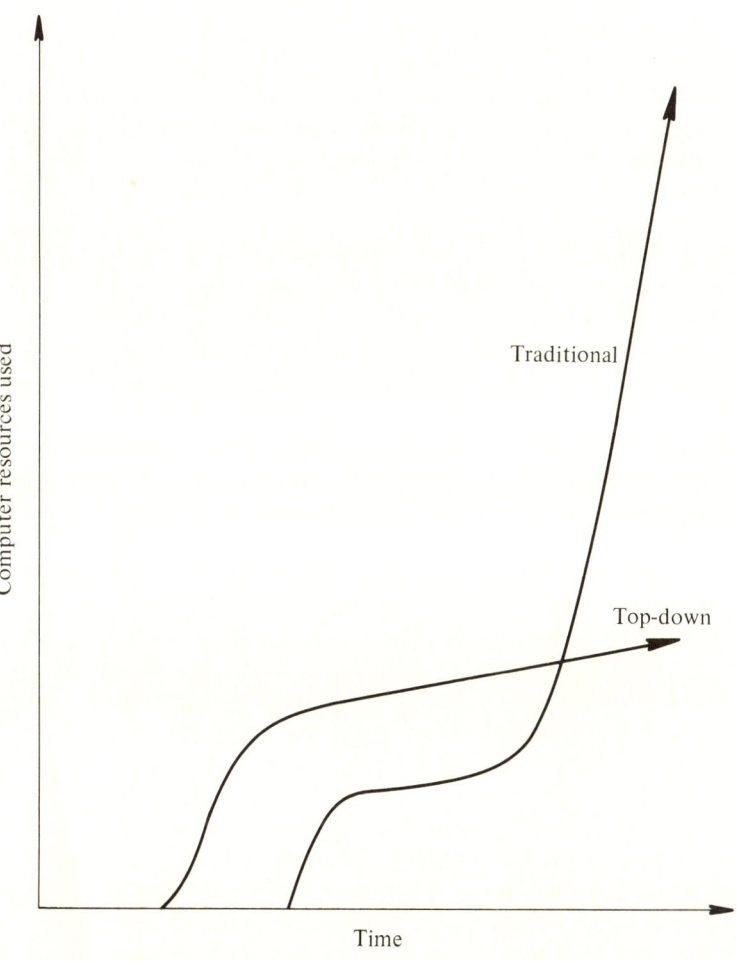

Figure 3.3:   Functional tree structure

Currently, top-down development organizes the system into a functional tree structure of program segments. This new design approach is subtle and requires a thorough study of program requirements before coding.

Hierarchical Tree Structure

As indicated above, top-down design and development begins with the definition of functional requirements and then proceeds to develop a high-level control segment followed by development and integration of detailed, lower level segments. Individual segments are incorporated as they are coded which requires continual testing of logical interfaces and linkages prior to program completion. In order to accomplish this, a software product is organized into a functional hierarchical structure of program segments as illustrated in Figure 3.3. The highest level of control logic is represented by the top segment. From this level, operation passes to lower level segments, and for each successive level the process is repeated until all functional specifications of the system are satisfied with a detailed program segment. During early development phases, dummy code is temporarily substituted for undeveloped segments. These required dummy segments, referred to as *dummy stubs*, are included during a test run. As actual program segments are developed, they are inserted as production code. Thus, the segments are fully integrated at each level and their logic procedures tested before coding begins at the next lower level. At each level, standardized code structure enhances the development by minimizing the amount of branching. Each segment is logically

Figure 3.4: Typical program function

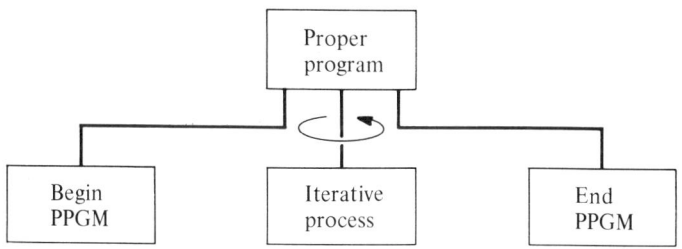

entered at the top and exited at the bottom. In addition, unconditional branching within a module is minimized or eliminated.

LOGIC DESIGN

While many sources give excellent advice on how to design, optimize, and evaluate functional segments, the primary problem of how to construct the basic control structure of a program is usually left as a creative exercise for programmers to solve on their own. This section is concerned with these basic control structures and factors which should be considered during their evolution. The basics of elementary top-down design can be reduced to two simple postulates that have widespread application in structuring programming problems.

Postulate 1: Every proper program function has a single beginning and a single ending. Implementing code always logically enters at the top and exits at the bottom. The schematic program structure shown in Figure 3.4 illustrates that a typical program has three basic functional elements: a beginning, an iterative process, and an ending.

Postulate 2: Top-down program design is achieved whenever that design reflects the required structure of input and output, plus all defined manipulative logic.

In most program design situations, only the general inputs and desired outputs are given. Thus, a logical starting point for translating functional requirements is to organize the process into a functional tree structure as in Figure 3.3. Program segments can typically be defined which will result in initial program logic specifications. Each one of these functional segments may then be further broken into three fundamental elements: input, process, and output, as shown in Figure 3.5.

Figure 3.5: Development of a functional hierarchical segment

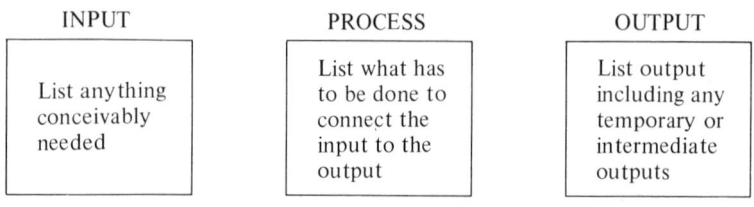

Input/Output Requirements

In any segment, the method by which I/O requirements are structured subtly determines program storage requirements. Basic logic designs which illustrate various I/O philosophies are shown in figures 3.6 through 3.10. Note that each design handles its I/O requirements uniquely. In the basic design (Figure 3.6), a single input is processed and the result is a single output. However, as I/O logic design becomes more complex, the determination of process structure becomes more critical since the programmer must establish how to maintain the required inputs and outputs. In general, four major options are available, each being an arrangement of array and nonarray capabilities. If program design does not utilize an array logic representation, then an interactive logic process with record I/O control is used (Figure 3.7). However, assuming adequate internal storage is available, the programmer may utilize arrays to reduce the total coding effort required to process large volumes of data. Here the logic design

Figure 3.6: Basic logic design

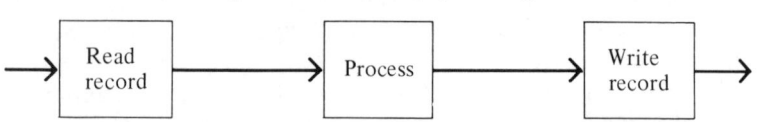

Figure 3.7: Iterative logic design with nonarray I/O control

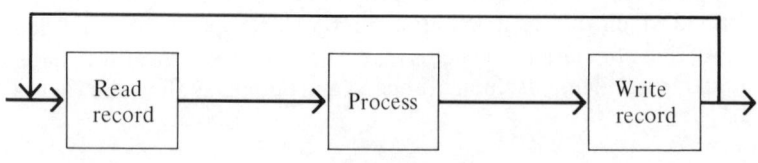

Figure 3.8: Iterative logic design with nonarray-input and array-output control

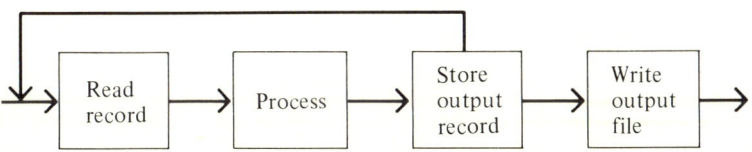

Figure 3.9: Iterative logic design with array-input and nonarray-output control

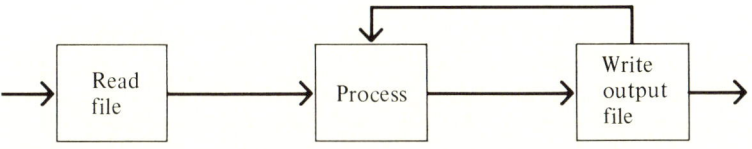

Figure 3.10: Iterative logic design with array I/O control

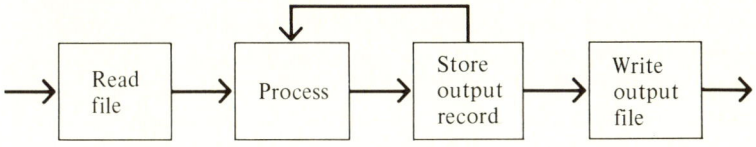

may incorporate internal storage for all input, for all output, or for both requirements. These concepts are illustrated in figures 3.8, 3.9, and 3.10, respectively.

Data Specifications

After general input and output requirements are determined, an examination of data specification should be performed. For example, PL/I provides the user with great data flexibility. The data may be handled as scalar, character, structures, arrays, or a combination of each. In addition, each specification may also contain the four characteristics of scale, base, mode, and precision. The overall choices of variable design options are shown schematically in Figure 3.11. Other high-level languages may not provide this degree of programming flexibility; however, the problem of data specification must always be considered in the design.

Figure 3.11: PL/I data specification options

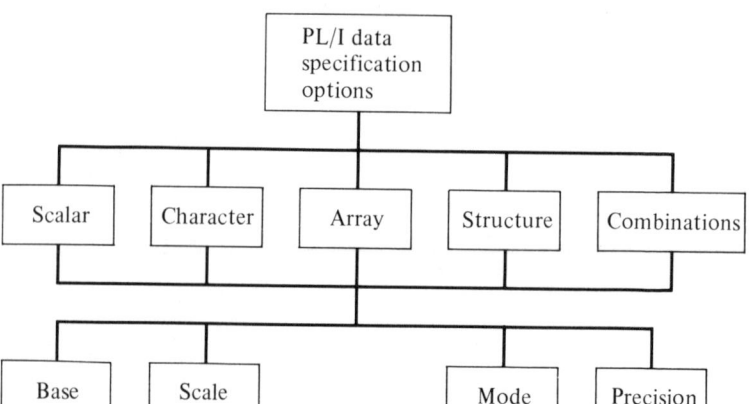

Logic Development

After major inputs and outputs have been established and data specifications defined, the individual logic processes may be developed. By utilizing the above I/O design considerations, a functional hierarchical design can be developed. Starting with three boxes labeled INPUT, PROCESS, and OUTPUT, needed input, required operations, and output are detailed (see Figure 3.5). As items are listed in the boxes, general overall logic requirements are defined and major functions identified. In addition, temporary and/or intermediate output is determined. Using the number aggregation example shown in chapter 2, one possible design solution is illustrated in Figure 3.12.

Note that as the items listed are connected by arrows, logical relations between elements are defined. The three items listed in the PROCESS box represent the major logical functions to be developed. In addition, process item one and the required output indicate an input/output logical design consisting of scalar I/O control, although, optionally, the input file might also be logically structured as an array. The result of these design decisions produces the first two layers of a functional top-down program design as shown in Figure 3.13. For a more complex problem, the next step would consist of repeating the above procedure for each major function using the three steps of input, process, and output. This procedure would then be continued for each major logical segment until a singular codable design module is reached.

Figure 3.12: Development of a functional hierarchical design for the summation of ten numbers

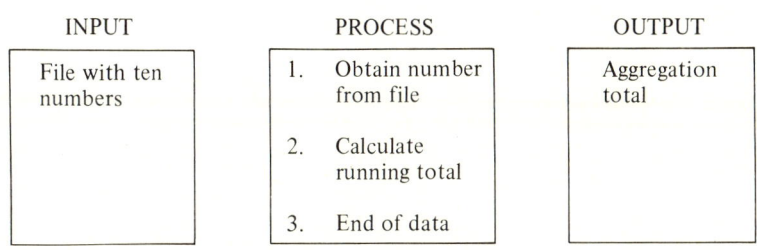

TOP-DOWN DESIGN CONSTRUCTS

Structured program code constructs are used as the key implementation elements of top-down design. In order to properly control segment execution three types of logic constructs are required. These are sequential, conditional, and iteration (repeated execution). A logic function in top-down design can be represented by a tree structure as shown in Figure 3.14. Here the segments A, B, C, and D are shown to be components necessary to accomplish some desired programming objective. The solid lines indicate that the segments A, B, C, and D are each executed by the main program control logic, in sequence from left to right. Thus, no conditional execution or iteration is specified.

Insertion of conditional symbols (◊), as shown in Figure 3.15, indicates that segments B and C may *not* be called in the sequence A, (B or C), D. The emphasis in preliminary top-down design is on broad

Figure 3.13: Functional top-down design

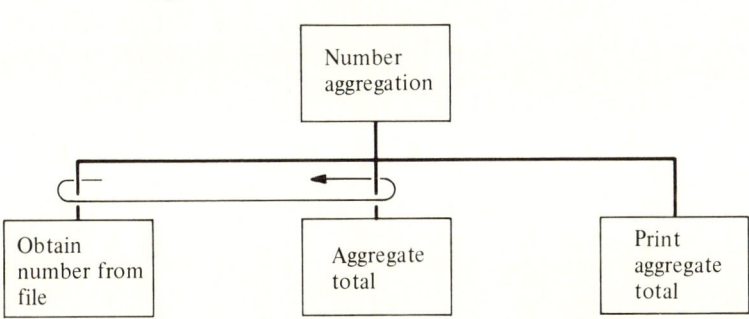

Figure 3.14: Sequential logic schematic

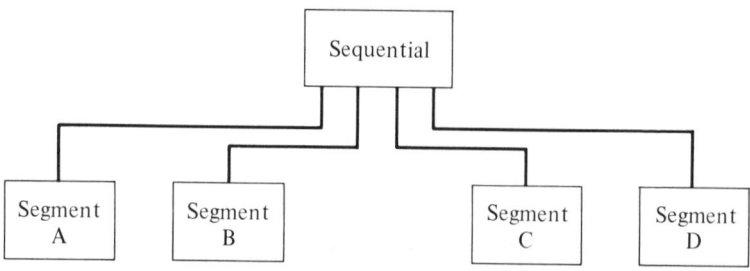

control structure and not the exact details of how conditional operations are to be implemented. Initial efforts focus on *identification* of the overall requirements. Subsequent research will further define the specific method of implementation. The conditional logic design illustrated in Figure 3.15 describes the top-down design equivalent to a conditional construct.

In Figure 3.16, the circular arrow above segment B indicates a repetitive execution of modules C and D. Again, the details of iterative loop termination are not required at this point. The key concern is the *recognition* that a module is to be executed repetitively. Later, attention will be paid to the specifics regarding control conditions. The repetitive logic schematic of Figure 3.16 is equivalent to an iterative (DOWHILE and DOUNTIL) construct.

Each of the top-down design figures shown thus far represents a basic control structure composed of separate logic segments. These segments or functions will eventually be expanded to contain the detail code logic necessary to implement their function; therefore, it is very important to understand how they are defined. One frequently

Figure 3.15: Conditional logic schematic

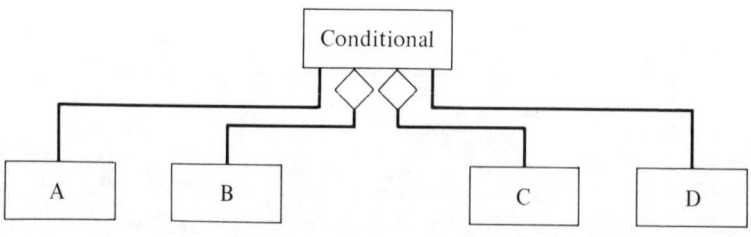

Figure 3.16: Iterative construct

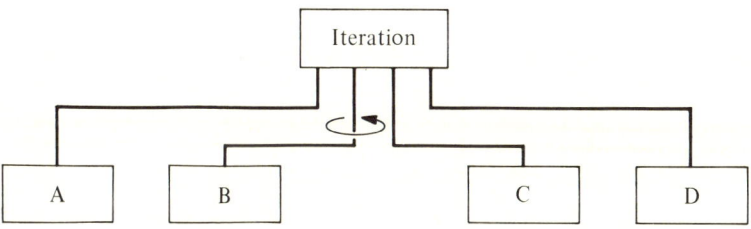

misunderstood concept is what constitutes a function or segment. It is viewed here as a set of logical instruction to implement a desired function. Whenever a segment is executed, the algorithm that manipulates an input into some desired output is the logical function of a segment. In top-down program design, a segment should be viewed initially as a black box. It doesn't matter *how* the segment performs its function. In fact it doesn't matter whether the segment accomplishes the function entirely internally, or if it requires other segments to accomplish its goal. Understanding of this concept is crucial to developing skills in top-down program design.

SUMMARY

Traditionally, programs were not designed. Instead, they evolved from crude flowcharts and hastily assembled code. Usually, the original design was quickly sketched in flowchart form to satisfy management's documentation requirement; however, this initial design slowly and subtly eroded when attacked by the "hard" facts of coding and debugging. Uncontrolled branching soon reduced the program to an incomprehensible mass of spaghetti logic. This is the "traditional" approach at its worst. In contrast to this, top-down program design is a hierarchical and evolutionary approach which enables the designer to proceed logically from the broad general aspects of program function to detailed code in an orderly manner. It is a philosophy as well as a methodology.

There are several reasons for the rapid acceptance of the top-down philosophy for program design. Experience has shown that past design methodologies have been accompanied by upward spiraling error rates and manpower costs. In an attempt to isolate problem areas with these design efforts, the following issues have arisen as significant elements hindering successful program management in a traditional design environment:

1. Unique programmer creativity results in logic structures which are difficult for other programmers to interpret.
2. Data definition and functional interrelationships are often inconsistent among common sets of programs.
3. Individual segment testing uncovers clerical errors in coding and logic errors in design after considerable cost is incurred.
4. Subsequent integration of logic segments reveals problems resulting from discrete module definition and coding which are often costly to patch or repair.

Researchers had these particular problem areas in mind as the top-down design philosophy emerged in the mid-seventies. However, if the top-down approach is to be a feasible alternative to attack the increasing cost of software development, there are several areas that must be examined before it can be accepted as a valid aid in software management. In general, the areas necessary to be understood are both procedural and behavioral in nature, and they can be broadly categorized as follows:

1. Behavioral resistance from programmers to standardization
2. Possible inefficiencies caused by redundancy of program code inherent in structured approach
3. Execution efficiency of structured code
4. Suitability of approach to high-level languages currently in use
5. Methods for teaching new techniques for program logic development
6. Workable procedures for generating initial functional requirements and translating them into program logic specifications

However, given the underlying theme that long-range maintenance of programs is becoming too costly, a significant effort to reduce cost must be made. Top-down design is a technique which, when properly applied, can decrease initial software cost and reduce the typical confusion caused by maintenance needs. From this introduction we can now begin to explore some of the more tangible activities involved in the software production process.

Top-Down Design

Review Questions

3.1 Discuss the differences between traditional and top-down design approaches.

3.2 What is the function of dummy stubs in top-down design?

3.3 What are the basic top-down design constructs? Do they provide the designer with the means to express all logical relationships? Discuss.

3.4 How is data specification handled in COBOL and FORTRAN?

3.5 What are the advantages of a hierarchical tree structure? What are the disadvantages?

3.6 Express the input/output control schematics as hierarchical tree structures.

3.7 What happens when a major function within a PROCESS box is composed of several functions?

3.8 Why might programmers resist the concept of top-down design?

3.9 In chapter 2 you were asked to use Warnier diagrams to develop logic to find the maximum and minimum numbers from a series of numbers. Solve the same problem using a hierarchical tree structure.

Notes and References

1. Myers, Glenford J. *Composite/Structured Design*. New York: Van Nostrand Reinhold Company, 1978.
2. _____. *Reliable Software Through Composite Design*. New York: Petrocelli/Charter, 1975.
3. Yourdan, Edward. *Techniques of Program Structure and Design*. Englewood Cliffs, N.J.: Prentice-Hall, Inc., 1975.
4. Yourdan, E. and Constantine, L.L. *Structured Design*. New York: Yourdan, Inc., 1975.
5. Stevens, W.P.; Myers, G.J.; and Constantine, L.L. "Structured Design." *IBM Systems Journal*, vol. 13, 1974, pp. 115–139.

4
STRUCTURED PROGRAMMING

The term "structured programming" connotes a recently developed approach for developing computer programs that can be written with a high degree of structure, thus facilitating ease of testing, maintenance, and modification. Development of this concept has evolved in conjunction with the top-down concept. As we now look at the design process, top-down development can be interpreted as both a macro and micro design philosophy. Chapter 3 outlined the development of logic functions, first at the system (macro) level, and later to successively lower level logical units. This chapter now looks at the coding (micro) level, where the concept of top-down design implies that program logic is to be constructed with standardized logic and controlled branching such that resulting code can be read from top to bottom.

The first attempt to formally define structured programming was developed in Europe by E.W. Dijkstra and C.A.R. Hoare in 1972. In summarizing their already classic work, the authors recognized the necessary code elements and rules to implement structured programming, but unfortunately ALGOL (used primarily in Europe) was the language used to develop this work. Programmers in this country were then faced with the problem of translation. Thus the past few years have primarily been spent in testing the structured programming approach and translating these concepts to languages which are more familiar.

Implementation of the structured programming philosophy requires two major activities: (1) functional system specifications which are compatible with the approach; (2) a rigorous coding discipline which minimizes logic construction variability. Neither of these elements is easy to achieve, and traditional language education has ingrained the "spaghetti" logic structure approach to coding.

Figure 4.1: Sequential operation

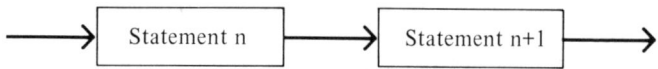

Figure 4.2: Conditional branch to one of two operations (IFTHENELSE)

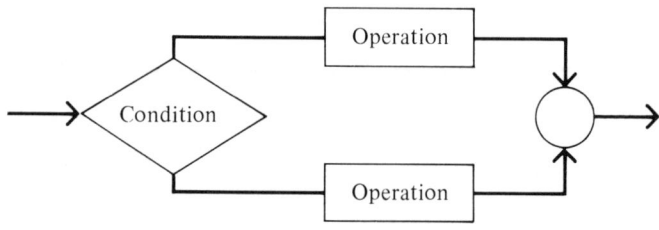

LOGIC CONSTRUCTS

In order to achieve technical standardization, basic logic constructs are utilized to control branching within each program logic module. These constructs limit the representation of complex problems by integrating and nesting a small number of standard logic structures; additionally, they require certain other coding practices to improve readability. The legitimate code blocks using structured programming theory are as follows:

1. SEQUENCE—a sequence of two or more operations (Figure 4.1)
2. IFTHENELSE—a conditional branch to one of two operations (Figure 4.2)
3. DOWHILE—repeating an operation while a condition is true (Figure 4.3)

Figure 4.3: Repeating an operation if a condition is true (DOWHILE)

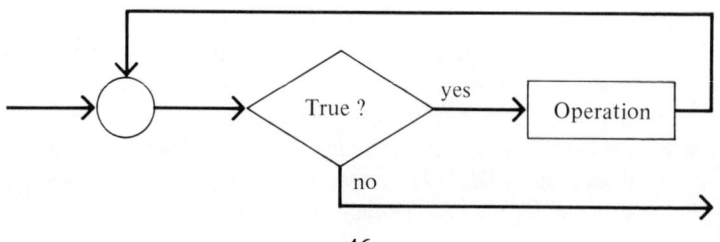

Structured Programming

Figure 4.4: Repeating an operation if a condition is true (DOUNTIL)

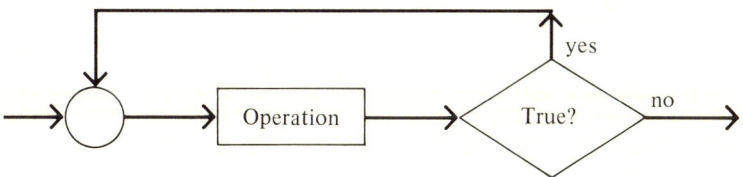

4. DOUNTIL—a conditional branch back to repeat an operation if a condition is true (Figure 4.4)
5. CASE—a conditional multibranch structure with one common exit point (Figure 4.5)

This basic set of logic structures is a practical extension of Bohm and Jacopini's original form, which proved theoretically that any problem can be broken down into small subproblems whose equivalent form can be expressed with only the first three logic types described above. However, from a practical coding viewpoint, all five logic types outlined above facilitate the process without destroying its basic intent. This approach to design decomposes the problem into multiple segments of logic of approximately sixty statements each, or one page, and uncontrolled branching is restricted. For example, branching is not allowed backward into the program and is used only to

Figure 4.5: Conditional multibranch structure (CASE)

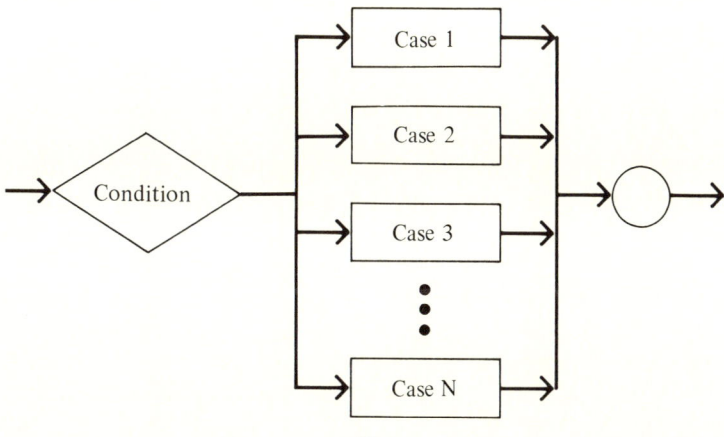

transfer control down to the next logical level (even this usage is frowned upon by the purist). In this manner, logic flow is more tightly restricted and easier to follow, plus readability and debugging are enhanced.

STRUCTURED PROGRAMMING WITH HIGH-LEVEL LANGUAGES

Early encounters with structured programming led many to believe that the approach meant "GOTO-less." In theory this is true; however, since the most commonly used languages were designed before this concept was recognized, such a restriction is often not practical. One overdue effect of this approach is the stark realization that none of our most popular languages is ideally suited for structure, yet we will show that each can be used in varying degrees to fulfill the spirit of the approach. In any language, the resulting code should be much improved when compared to traditional program design. If we accept the fact that structured programming is a way of achieving software management, then we must examine its usefulness with the three major languages used today—PL/I, FORTRAN and COBOL.

Implementation of structured program design and its associated code structures can be achieved in each of the above major languages. To accomplish this, symbols c and p are introduced:

c represents any valid Boolean expression (e.g., a logical condition)

p represents any code sequence and may contain sequences of operations, other basic coding structures, or any combination of these

Using this syntactical construction, the following four sections illustrate the actual skeleton source code necessary to implement the structured programming constructs. Since the SEQUENCE construct simply means sequential execution of statements, no example of this is shown here as all three languages normally operate in this fashion.

IFTHENELSE Logic Structure

In this construct, a process (p_1) is executed when a condition (c) is true, while another process (p_2) is executed when the condition is false. This construct poses a problem for FORTRAN in that the IFTHENELSE is not part of the available code. Therefore, in order to implement it, GOTO's must be used. A skeleton code for this is shown below:

```
C       FORTRAN IFTHENELSE CONSTRUCT
        IF (c) GO TO 10
                p₂
        GO TO 20
10              p₁
20      CONTINUE
```

However, this does not indict the language since the top-down and one entry-one exit requirements are still met.

In both PL/I and COBOL, the IFTHENELSE construct can be accomplished directly since both languages contain the construct as part of the language:

```
/* PL/I IFTHENELSE CONSTRUCT */
IF c THEN DO;
    p₁
        END;
ELSE DO;
    p₂
        END;
```

```
*COBOL IFTHENELSE CONSTRUCT
    IF c THEN
        p₁
    ELSE
        p₂.
```

DOWHILE Logic Structure

In considering this SP construct, two situations must be looked at. The first is to repeat a process while a condition exists. The second is to use the construct in conjunction with indexing.

Iterative DOWHILE. Both PL/I and COBOL can easily incorporate the first situation as shown below:

```
/* PL/I DOWHILE CONSTRUCT */
DO WHILE (c);
    p
END;
```

*COBOL DOWHILE CONSTRUCT
          PERFORM PARA THRU PARA-END UNTIL c.
          GO TO NEXT-PARA.
     PARA.
          p
     PARA-END. EXIT.
     NEXT-PARA.

The FORTRAN implementation of the iterative DOWHILE is illustrated by the following statements:

```
      C     FORTRAN DOWHILE CONSTRUCT
            LOGICAL COND, NOTCOND
            COND = (Boolean expression)
            NOTCOND = .NOT.COND
      C     BEGIN DOWHILE
      20    IF (NOTCOND) GO TO 40
               p
            GO TO 20
      C     END DOWHILE
      40    CONTINUE
```

From the previous code, the FORTRAN DOWHILE construct violates the structured programming requirement that the code not contain GO(back)TO. While PL/I and COBOL can implement the DOWHILE, PL/I is the only language that handles it directly.

DOWHILE with Indexing. In PL/I and COBOL, the DOWHILE with indexing is constructed with the following code structures:

    /* PL/I DOWHILE CONSTRUCT WITH INDEXING */
    DO index WHILE (c);
        p
    END;

    *COBOL DOWHILE CONSTRUCT WITH INDEXING
         PERFORM PARA THRU PARA-END index UNTIL c.
         GO TO NEXT-PARA.
    PARA.
         p
    PARA-END. EXIT.
    NEXT-PARA.

Structured Programming

An explicit example for the PL/I DO statement is illustrated below:

DO I = −10 TO 0 BY 1 WHILE (X > 4);

An equivalent COBOL PERFORM verb to supply indexing is:

PERFORM PARA THRU PARA-END VARYING I FROM 10 BY −1 UNTIL I = 0 OR X > 4.

The following is FORTRAN source code illustrating the DOWHILE structure:

```
      C     FORTRAN DOWHILE WITH INDEXING
            LOGICAL COND, NOTCOND
            I = 10
            COND = (Boolean expression)
            NOTCOND = NOT.COND
      C     BEGIN DOWHILE
      20    I = I − 1
            IF (I.EQ.0) GO TO 40
            IF (NOTCOND) GO TO 40
               p
            GO TO 20
      C     END DOWHILE
      40    CONTINUE
```

These examples illustrate that the only languages which employ a statement allowing for an indexing condition and a Boolean condition are COBOL and PL/I. Again, FORTRAN falls short because a GO(back)TO must be used with an additional IF statement.

DOUNTIL Logic Structure

The DOUNTIL construct must be considered with and without indexing. Without indexing, only certain PL/I compilers allow for direct implementation of this construct. However, with minor modifications both PL/I and COBOL can be adapted, while FORTRAN again exhibits the same problems it did with the DOWHILE. This can be seen with the following code segments:

```
/* PL/I DOUNTIL CONSTRUCT */
DECLARE DONE BIT(1) INIT('0'B);
/* 0 BINARY EQUATES TO FALSE */
/* 1 BINARY EQUATES TO TRUE */
```

A Primer on Structured Program Design

```
        DO WHILE (DONE);
            p
            IF c THEN DONE = '1'B;
            /* NOT CONDITION */
        END;
```

```
*COBOL DOUNTIL CONSTRUCT
77 DONE PIC 9 VALUE 0.
    •
    •
    •
    •
PARA-A.
    p
    IF c THEN MOVE 1 TO DONE.
PARA-A-END. EXIT.
PARA-B.
    PERFORM PARA-A THRU PARA-A-END UNTIL DONE = 1.
```

```
            C       FORTRAN DOUNTIL CONSTRUCT
                    LOGICAL NOTDONE
                    NOTDONE = .TRUE.
            C       BEGIN DOUNTIL
            10      CONTINUE
                        p
                    IF c NOTDONE = .FALSE.
                    IF (NOTDONE) GO TO 10
            C       END DOUNTIL
```

It must be noted that COBOL actually allows for the DOUNTIL easier than FORTRAN and some versions of PL/I. FORTRAN's failure is again the need for a GO(back)TO. COBOL and PL/I can both adjust to the indexed DOUNTIL. The only changes would be the addition of the VARYING verb in the COBOL PERFORM and the addition of indexing to the PL/I DOWHILE.

CASE Logic Structure

The fifth construct can be accomplished similarly in all three languages. They use a form of the GOTO to accomplish the case construct. Coded examples of these logic structures are shown with the following source code:

```
          C    FORTRAN CASE CONSTRUCT
               GO TO (10, 20, 30, 40) CASE
          C    ERROR CASE
                    pe
               GO TO 50
          C    CASE 1 LOGIC
          10        p1
               GO TO 50
          C    CASE 2 LOGIC
          20        p2
               GO TO 50
          C    CASE 3 LOGIC
          30        p3
               GO TO 50
          C    CASE 4 LOGIC
          40        p4
          50   CONTINUE
```

```
*COBOL CASE CONSTRUCT
     GO TO (CASE-1, CASE-2, CASE-3, CASE-4)
          DEPENDING ON CASE.
CASE-ERROR.
     pe
     GO TO END-CASE.
*CASE 1 LOGIC
     p1
     GO TO END-CASE.
*CASE 2 LOGIC
     p2
     GO TO END-CASE.
*CASE 3 LOGIC
     p3
     GO TO END-CASE.
*CASE 4 LOGIC
     p4
     GO TO END-CASE.
END-CASE.
```

```
/* PL/I CASE CONSTRUCT - STANDARD VERSION */
/* FOUR VALID CASES AND ONE ERROR */
DECLARE CASE(5) LABEL;
CASE __INDEX = (integer created from program logic)
/* ERROR TEST */
IF CASE __INDEX < 1 | CASE __INDEX > 4 THEN
    CASE __INDEX = 5;
GO TO CASE(CASE __INDEX);
/* CASE 1 LOGIC */
CASE(1):
    p₁
GO TO END__CASE;
/* CASE 2 LOGIC */
CASE (2):
    p₂
GO TO END__CASE;
/* CASE 3 LOGIC */
CASE(3):
    p₃
GO TO END__CASE;
/* CASE 4 LOGIC */
CASE(4):
    p₄
GO TO END__CASE;
/* ERROR MESSAGE */
CASE(5):
    pₑ
END__CASE:
```

Even though each requires the use of a GOTO, it must be remembered that pragmatic structured programming does not mean "GOTO-less." All three sets of code illustrated here do adhere to the in-line code requirement. FORTRAN and COBOL allow for an error case directly, while PL/I must check for an error with an IF statement. In addition, enhanced PL/I versions contain commands for direct implementation of the CASE construct. The following code illustrates the CASE stucture for the PL/I Optimizer and Checkout compilers:

```
.* PL/I CASE CONSTRUCT-OPTIMIZER AND CHECKOUT
COMPILERS */
SELECT;
```

Structured Programming

```
WHEN c₁ DO;         /* CASE 1 LOGIC */
    p₁
        END;

WHEN c₂ DO;         /* CASE 2 LOGIC */
    p₂
        END;
WHEN c₃ DO;         /* CASE 3 LOGIC */
    p₃
        END;
OTHERWISE DO;       /* ERROR TEST */
    pe
        END;
```

Comparative Evaluation

The previous sections explored the necessary coding mechanics of structured programming through the construction of five basic logic blocks. Specific coding examples using FORTRAN, COBOL, and PL/I were shown for each block type to illustrate the elementary mechanics of each. Table 4.1 summarizes the relative ease by which each block

Table 4.1. Language Compatibility with Structured Programming

	Basic Block	*Language*			
	Type	COBOL ANSI	PL/I Optimizer	FORTRAN (G level)	
1	Sequential	1	1	1	
2	IFTHENELSE	1	1	2	required
3	DOWHILE				blocks
	Repeat	2	1	4	
	Index	1	1	3	
4	DOUNTIL				
	Repeat	1	1	4	extended
	Index	2	1	2	blocks
5	CASE	2	2	2	

Legend (Note: Figure 4.6 also)
1 = Easily incorporated
2 = Fairly easy to incorporate
3 = Possible to incorporate, but not considered easy
4 = Difficult to incorporate. Does not adequately serve the needs of structured programming
5 = Impossible to incorporate

Figure 4.6: Comparison of languages for basic logic constructs

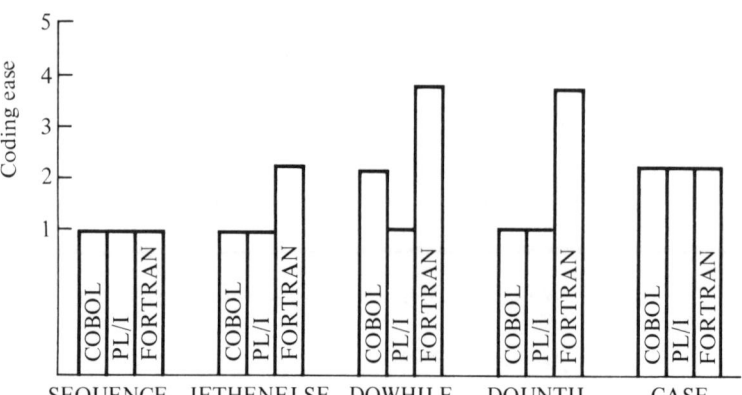

type can be implemented using normal language options. No formal attempt has been made to evaluate a particular language's compatibility to structured programming principles, although random specific criticisms do appear. The first three block types are required by structured theory, while types four and five are modifications added for flexibility. These evaluations are addressed only to the specific language syntax required to implement structured programming. There are obviously other factors which must be considered in the language selection process.

Another view of the compatibility of each language to the structured logic constructs is shown in Figure 4.6. Be aware that the rankings shown represent the authors' judgment and are not necessarily a concensus. When evaluating structured programming requirements, it seems safe to conclude that FORTRAN is a poor choice as a high-level language. As PL/I achieves broader compatibility with various hardware manufacturers, then its use should increase at the expense of FORTRAN. A block orientation and ability to process logic macros increase the attractiveness of PL/I. The major attractiveness of COBOL is its general level of hardware compatibility and a well-understood structure of its own. Commercial data processing operations are strongly attached to COBOL and this does not seem likely to change. Regardless of the direction taken, the recognition of a theoretical logic structure which can be practically applied will undoubtedly have a major impact on future problem-solving approaches using computers.

SUPPLEMENTARY CODING STANDARDS

The development of software systems has been recognized as first and foremost a management problem. With the increasing emphasis on software cost and long-range maintenance of programs, standards are being developed which go beyond those of structured programming. These standards are the foundation for controlling entire software development efforts. One significant phase of development paralleling structured programming definition incorporates a number of standards for writing source code.

There is great value in utilizing a good set of coding standards. First, management as well as other programmers may more easily read and understand the structure of programs written by others. Second, development of coding techniques reduces the complexity of programs. Finally, standards provide management with the guidelines to control the overall programming effort.

In Table 4.2, sample program standards are presented which are essential for development of quality programs. The primary objectives of these standards are to enhance the readability of code and to facilitate maintenance of these programs. As a sample of a larger effort to promote high-quality programs, this set can be expanded to encompass features unique to the different languages. The issue of program standards is complex; nevertheless, the standards given here are only broad guidelines for assisting managers and programmers in developing usable standards.

CONCLUSION

In the current revolution of program design, attention is increasingly focused on the human aspects of program development. Development, testing, and maintenance costs are now recognized as keys to reduced processing cost. As a result, techniques such as structured programming and its associated coding standards are being utilized to provide more control over the programming effort. This chapter addressed the micro portion of structured design. The objective was to present an overview of the associated construction of source code for the most popular languages: COBOL, FORTRAN, and PL/I. Obviously, the adoption of good program conventions is only a part of the process required to produce high-quality software. However, the issues presented here are important elements of the current trend toward increased standardization.

Table 4.2. Program Standards Essential for the Development of Quality Programs

Item Number	General Rule	Function
1	For each installation and for each application, there should be an adopted set of standard user-defined words (e.g., variable names and logic segments).	Naming conventions reduce time spent on devising mnemonic variables. A published data dictionary promotes program readability and prevents confusion.
2	The names of all persons who have written or changed code should be included in every program, or included in technical documentation.	Promotes quick access to those responsible for a program.
3	After a program is in a production environment, comment lines describing each modification and a version number should be added before every recompilation.	Provides documentation containing the nature and reason for the change.
4	The maximum length of a functional segment is 50–100 lines of executable code.	Limits the complexity of each segment.
5	Program termination may only occur in the last statement of the main segment of a program and logic flow is top to bottom.	Forces a complete use of a single-in, single-out control structure.
6	Nesting of IF constructs can be at most three levels deep. When more than this, the CASE construct should be used.	Eliminates confusing nesting constructs.
7	The values of control variables associated with indexing may not be modified by the performed process.	Prevents deceptive loop control errors.
8	Input-output operations should be isolated in separate processes and invoked by a CALL or PERFORM instruction.	Isolates the program elements which require external input-output.
9	Edit external data immediately after input to filter bad out of program logic.	Avoids using data values which can cause errors in subsequent processing.

Review Questions

4.1. What are the basic constructs utilized to control branching within program modules?

4.2. Draw a schematic to represent each basic logic construct.

4.3. Discuss how top-down programming facilitates program correctness.

4.4. Take an existing program and analyze its structure.

4.5. Discuss the adaptability of COBOL, FORTRAN, and PL/I to structured programming concepts.

4.6. Table 4.2 contains general program standards essential for the development of quality programs. Develop a supplementary list of standards which applies to your design environment.

4.7. Take an existing program and modify the code to include structured program coding standards.

4.8. Write the structured programming code for the logic developed in chapter 3 for finding the maximum and minimum numbers from a series of numbers.

Notes and References

1. Bohm, C. and Jacopini, G. "Flow Diagram, Tuning Machine and Languages with Two Formation Rules." Communications of the ACM, May 1966, pp. 366–371.
2. Dahl, O.-J.; Dijkstra, E.W.; and Hoare, C.A.R. *Structured Programming.* New York: Academic Press, 1972.
3. McGowan, Clement L. and Kelly, John R. *Top-Down Structured Programming Techniques.* New York: Petrocelli/Charter, 1975.
4. Mills, H.D. *How to Write Correct Programs and Know It.* Gaitherburg, Md.: IBM Corporation FSC73-5008, 1973.
5. Yourdan, Edward. *How to Manage Structured Programming.* New York: Yourdan, Inc., 1976.

5
STRUCTURED DESIGN TECHNIQUES

At this point in our discussion, we have essentially reviewed the history and key theoretical concepts of structured programming. It is now time to begin illustrating how these abstract notions can be implemented into high-quality systems and programs which will produce the advantages found by users of structured programming.

Chapters 5 and 6 are companion discussions. In chapter 5 we intend to look at the higher level issues of design structure, evolution, and communication. Chapter 6 will emphasize the implementation (coding) portion of the process. Collectively, the two intertwined subjects are intended to help the practitioner (designer or programmer) become more capable of producing structured systems and programs. HIPO (hierarchy plus input-process-output) charts will be used as a vehicle to describe the structured design process.

The traditional approach to systems design is for a user to indicate the desire for some new system capability. It then typically falls on the systems analyst to translate these vague requirements into a concrete, cost-effective information system. All too often the user is far removed from the specification process until a "final" working version is presented for approval and use. At this point two responses are all too frequent: "Oh no, that is not what I meant!" "Hey, this is great! How about adding a couple of bells and whistles [e.g., enhancements] right there?" Both responses are bad news for the designer and both reflect a lack of user-designer communication. Realistically, there must be more perfect communication in regard to function before coding; however, this really represents only the top of the traditional design iceburg. We need to know more about why programs and information systems fail all too often after they are in production. Also, why are many users unhappy with existing levels of performance (e.g., turnaround or response time)?

Assuming away the possibility of designer incompetence, there are several reasons why approximately one-third of a data processing budget is spent on reworking existing software. Much of this results from errors in the original systems design or sloppy implementation of a reasonably good design. More specifically, errors can usually be traced back to improper requirements specifications in one of the following areas:

1. Output function—doesn't work as specified or perceived
2. Input formats—bad data not filtered out before entering the program
3. Logical processes—legitimate data produces erroneous results
4. Stored data--needed historical results not saved or difficult to retrieve
5. Access controls—users illegally modify stored data
6. Testing specifications—logic errors not uncovered during testing because of poor procedures
7. Documentation—an apparently simple modification causes catastrophic results elsewhere in the system

There are other scenarios which could shed additional light on the existing problems of systems design, but they are too voluminous to fully describe here. We can say, however, that the results of a system or program design activity are all too often characterized by the following:

1. Documentation for users and maintainers is lacking or absent (e.g., many organizations feel that the source code is good enough).
2. An examination of existing programs reveals that the programs are filled with GOTO instructions and vague structure, so that if one connected all branching instructions the resulting listing would look like a spaghetti bowl.
3. Design projects are late in delivery, and users are generally not happy with what they do get.
4. Productivity of programmers is apparently low (approximately four to twelve lines of code per day), although this is certainly a weak measure of output value.

Structured Design Techniques

5. Overall software development cost is spiraling upward to between 60% and 90% of the DP budget.

Obviously, no single approach or control tool will solve all of these problems. However, it is our bias that a tool such as HIPO can be a *vehicle* to aid in developing better systems which avoid many of the problems summarized above. As we proceed forward there are four goals of this discussion. These are:

1. Improve reader understanding of structured (functional) system and program design techniques.
2. Introduce key issues and concepts of structured design.
3. Outline a procedure for structured system development procedures.
4. Generally describe mechanics of the HIPO design tool and its relationship to the above.

SYSTEM LIFE CYCLE

All systems whether they be biological, manual, or computerized evolve through a birth to death process. Also, every system is characterized by varying degrees of complexity. Basically, computerized systems must be managed through the following eight steps:

1. Requirements analysis and definition
2. System definition
3. System design
4. Program design
5. Detailed module design
6. Testing and acceptance
7. Documentation (user and technical)
8. Modification and maintenance

As indicated previously, a systems analyst transforms user requirements into a tangible set of specifications through an iterative process. Theoretically, after the specifications are developed *all* that is required is to code the necessary set of programs. In fact, the process is much more complex than this. Often the designer is too vague, omits a function, or improperly defines the process (algorithm) by which the data is to be manipulated. Step 6, testing, provides the last opportunity to identify flaws in the design. Later, after the system is in use even small changes require relatively large expenditures of effort and an increased amount of confusion.

Figure 5.1: Hierarchical view of a system

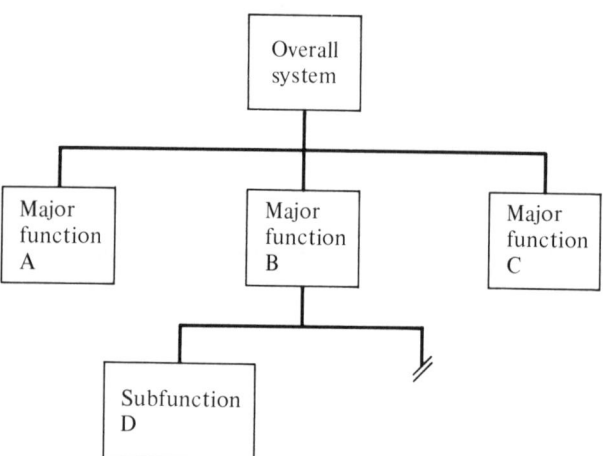

What emerges from this discussion is a realization that coding should not be started until design is finished and approved by the user. Once coding has begun, certain rules should be followed which segment, or structure, the logic. Possibly the following analogy makes the point: Only after creating a firm foundation can the temple become quite large and sturdy. Temples built in the swamp may tilt badly, or simply sink out of sight. Life cycle costs are greater if the design is not held firm, yet, typically, the user is in a hurry, thus the design step often gets cut short.

Information or other data processing systems are created to serve some broad function, therefore the design must define the function. Too often designs spend excessive time working on data structures and too little time on function. The rule is function first and data structure later. Conceptually, a system can be viewed as a hierarchy of functions as shown in Figure 5.1.

Initially, the system design should be concerned only with *what* it is to do, not *how*. Also, it is better to stretch technology in the beginning than later after a more simple design approach has to be modified. Think of future needs and not just the present. Successful companies grow and so do successful systems. Consider such things as potential items to be stored and future user environments. Once these considerations are made the first phases are restricted to major and

Structured Design Techniques

second-level functions in a top-down manner. As a design evolves the number of levels increases and so does the amount of detail. Eventually the *how* will have to be resolved, but hold off on this as long as possible since it will cloud otherwise clear design choices.

After the initial design phase, a general logical structure should exist which satisfies the user and designer. In addition, milestone-type tentative schedules are needed early, along with budget guidelines. Management should *not* be asked to buy something which is undefined, to be delivered at an unspecified time and at an undefined cost. Ludicrous as this sounds, many information system designs are begun with just such a vague understanding.

As the design progresses the hierarchy grows (essentially vertically) and the management development plan becomes more firm (e.g., time, cost, and performance goals). At this time such issues as subsystem phasing, hardware procurement, and training are woven into the overall plan. Also, we are ready to begin the coding, testing, and documentation phases of the life cycle.

FUNCTIONAL DESIGN CONCEPTS

As the design evolves it passes through the following stages of concern: system, subsystem, program, module, and logic segment. Figure 5.2 gives a skeleton view of the module level of detail which might be found in a hypothetical material control management information system. Note that at each level in the hierarchy we are defining increasingly more detailed events.

Traditionally, flowcharts have been used to schematically reflect design details, first systems flowcharts and later more detailed program logic flowcharts. These tools have stood the test of time and undoubtedly will endure, at least for the traditionalist. However, flowcharts are GOTO oriented and programs written from flowcharts reflect this orientation.* Since the goal here is to modularize and structure a design, we think a hierarchical tree diagram better lends itself to the task. Notice that nowhere in our future discussion do we show GOTO-type branching—simply top-down linkages of logical functions.

Each box in the hierarchical design represents an abstraction of some level of logical function. Finding a single descriptive name for all

---

*There is a new type of flowchart called structured flowchart which does not contain this deficiency.

A Primer on Structured Program Design

Figure 5.2: Module level of detail

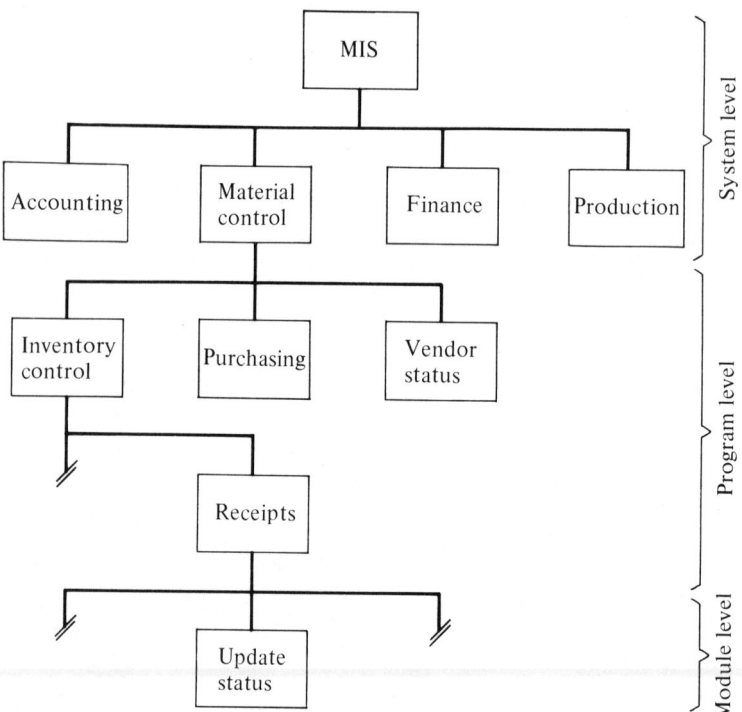

boxes in the tree structures is difficult, but let us call them *modules* for now. When implemented each module should have a single entry and single exit; it should also be of manageable size and contain only legitimate structured programming control flows. The preferred relationship among modules is for each to simply pass data to the next lowest level module[7]. This attribute is called *data coupling*. During the design phase various module interactions are specified by the design. The nature of these interactions is called *module coupling*. Simply stated, we mean what does one module need from another. If one module alters the code in another, then the two are tightly coupled. Good functional design attempts to maintain a well-defined structure for all modules with minimal coupling. Flow control is top to bottom and left to right, except when general-purpose modules are

Figure 5.3: Functional tree structure implementation

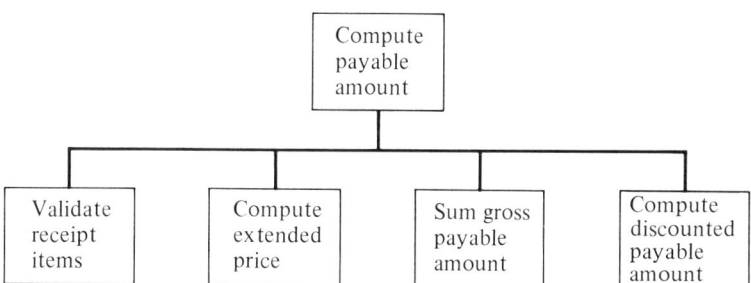

needed. In this case a call and return is clearly specified (more will be said of this later).

Stay provides a highly simplified example to demonstrate the process of functional design.[7] Suppose we are designing an accounts payable system and one high-level function is COMPUTE PAYABLE AMOUNT. Figure 5.3 shows this function reduced to its next lowest level of detail. Implementation of the logical structure is left to right, that is, the following PL/I code could implement the lower level of this structure assuming each of the modules is constructed as a procedure:

```
CALL VALIDATE_RECEIPT_ITEMS;
CALL COMPUTE_EXTENDED_PRICE;
CALL SUM_GROSS_PAYABLE_AMOUNT;
CALL COMPUTE_DISCOUNTED_PAYABLE;
```

Notice that the implementation of these functional models is left to right.

Closely associated with the hierarchical structure is the issue of functional process. This is generally defined in the beginning as a CRT screen layout or a paper report format. Thus user design specifications tend to be *output* oriented. Unfortunately, computer systems design is more closely aligned with inputs (tape formats, data base structures, etc.). Once described the data output becomes trivial. The common denominator between designer and user is then process. This linkage is shown schematically in Figure 5.4. A chart such as this one is often called an IPO diagram which is an acronym for input-process-output. When combined with a hierarchical design the result is HIPO. This overall approach epitomizes functional design.

Figure 5.4: Input-process-output design orientation

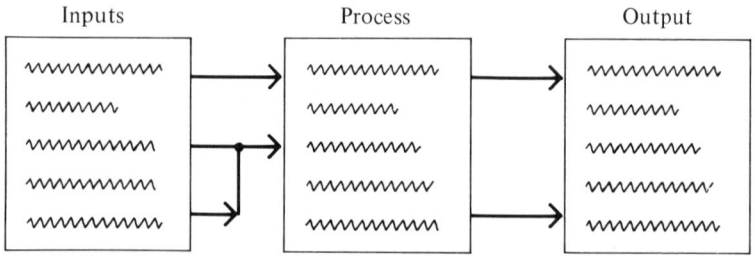

ROLE OF HIPO IN SYSTEMS DESIGN

Broadly defined, hierarchy plus input-process-output (HIPO) is a graphical procedure to accomplish functional system design through a series of iterative steps. In this section we will outline some of the concepts of this tool. When viewed collectively, HIPO is a graphical and functionally oriented tool intended to display *what* a system or program does and what data it uses and creates.[4] In this sense it is both a design and documentation tool. Throughout the HIPO exercise schematic design structures are used to focus on *what* a system does and *not how* the functions are performed.

There are three component parts of HIPO. These are:

1. Visual table of contents (VTOC) tree diagram
2. Overview diagram (IPO charts)
3. Detail diagram

Although each of these parts has a set of rules and charting conventions, we want to emphasize that there is no right way or wrong way to use HIPO. Its format should be somewhat standardized locally, but modified to fit individual design needs.

Visual Table of Contents Diagram

The VTOC serves as the overall organization of the design in much the same way as an organization chart broadly defines the functions within a business. Figure 5.5 illustrates the VTOC format. Basically, this form of overview should show a reader the structure of the system and give a reasonable impression of the system functions. Undoubtedly, the most valuable design function of the VTOC tree diagram is the numbering system. It is through this that various lower level design

Figure 5.5: Visual table of contents illustration

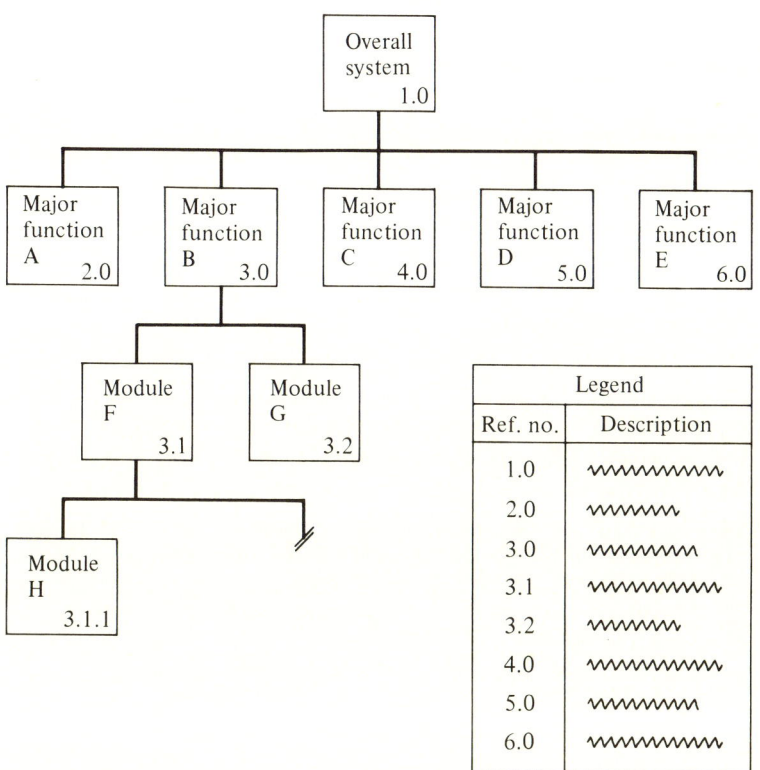

details can be referenced and located. This reference logic can be maintained in paper documentation, as well as subsequent source programs.

Overview Diagram

An overview diagram represents a generalized input-process-output (IPO) diagram describing what is done in each logical section identified in the VTOC. Figure 5.6a schematically illustrates the format of this chart. Note that at this point in the design no attempt is made to do anything other than enumerate the major logical segments to be accomplished to satisfy the function under study. Arrows at the top indicate flow of control from module 4.1; at termination, control is passed to module 4.2.

Figure 5.6: (a) Overview diagram example

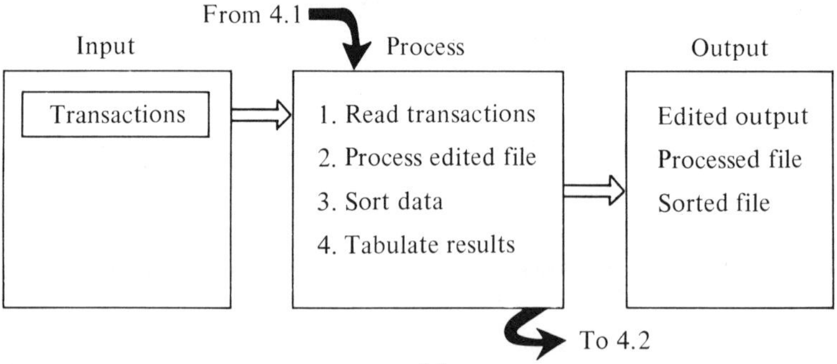

```
           From 4.1
  Input               Process                  Output
┌─────────────┐   ┌──────────────────────┐   ┌──────────────┐
│ Transactions│──▷│ 1. Read transactions │   │ Edited output│
│             │   │ 2. Process edited file│──▷│ Processed file│
│             │   │ 3. Sort data         │   │ Sorted file  │
│             │   │ 4. Tabulate results  │   │              │
└─────────────┘   └──────────────────────┘   └──────────────┘
                                    ↘ To 4.2
                        (a)
```

Detail Diagram

The purpose of detail diagrams is to further define the logical flow of data through selected portions of the system. Figure 5.6b details the logic further from our simple overview diagram. Notice from this example that there is some crude attempt to link input data to output files. Data flows can be represented by arrows depicting the following:

1. Solid (■──▶)—linkages from or to other portions of the design structure, or other utility programs.
2. Regular (▭─▷)—logical data flow between input or output data and indicated logic processes.
3. Dashed (- - - -▶)—(not illustrated here) a data reference such as a look-up table.

Implementation of HIPO

Designing a system using HIPO is consistent with the top-down philosophy. We have just seen introductory examples of the three basic HIPO descriptive charts. Now let us see how they relate to each other. The design begins as VTOC diagrams, then proceeds through overview diagrams, and finally to detail diagrams. Unfortunately, this process is not merely sequential, but rather it is iterative. By that we mean the design is refined, expanded, detailed, and then the process is repeated until the function under study is logically completed. This process is called *iterative refinement*. There is no prescribed manner to discover the major functions or segments of a system, but usually the top levels of the design represent major functional components of

Figure 5.6: (b) Detail diagram example

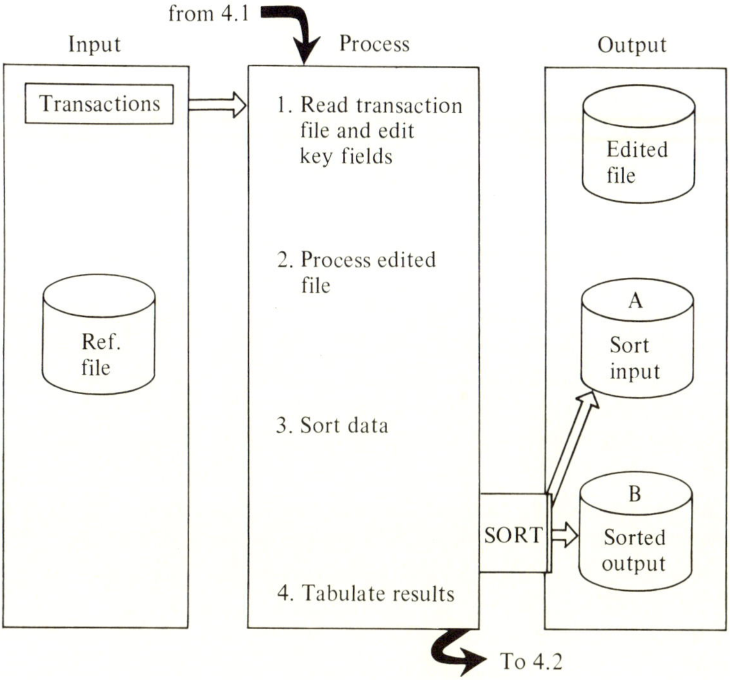

```
                from 4.1
    Input          ↓  Process                   Output
 ┌─────────────┬──────────────────────┬─────────────────┐
 │ Transactions│  1. Read transaction │                 │
 │             │     file and edit    │   Edited        │
 │             │     key fields       │   file          │
 │             │                      │                 │
 │             │  2. Process edited   │                 │
 │             │     file             │      A          │
 │   Ref.      │                      │   Sort          │
 │   file      │                      │   input         │
 │             │  3. Sort data        │                 │
 │             │                      │      B          │
 │             │                 SORT │   Sorted        │
 │             │                      │   output        │
 │             │  4. Tabulate results │                 │
 └─────────────┴──────────────────────┴─────────────────┘
                              ↳ To 4.2
```

Extended Description			
Process block	Procedure name	Pseudo-code diagram	VTOC ref.
1. Read transaction	READ_TRANS	READ4-1	4.1.1
2. Process edited file	PROCESS_FILE	EDIT4-2	4.1.2
3. Sort data	SORT	SORT4-3	4.1.3
4. Tabulate results	TAB_RESULTS	TAB4-4	4.1.4

(b)

the system. Each lower level then expands upon the level above. As the system design evolves, lower hierarchical levels become increasingly more detailed. Design of this logic basically starts with the specified outputs. From this it takes varying amounts of time to construct needed inputs and intermediate processes. It is our contention that design is still an art form, although less so than twenty years ago.

The design of modern information systems is not easy and becomes increasingly complex as on-line access and data base considerations are added to the problem. In any event the designer does not just sit down and draft the "grand design." It takes many months of effort to produce a reasonable design for a medium-sized system (25–50 thousand lines of source code). Katzan uses the schematic view of iterative refinement shown in Figure 5.7 to show this process.[4]

As we shall see in chapter 6, well-constructed VTOC and IPO diagrams will lead easily from the design through pseudo-coding to final code. This tangible transformation track is one major function that we are seeking. Even when loosely drawn, HIPO-type documentation serves as a communication tool to describe what the design is. Well-constructed HIPO diagrams serve all facets of the design and subsequent maintenance functions.

Jones describes the use of HIPO in developing system specifications at the John Hancock Mutual Life Insurance Company.[3] The effort described was successful in that the resulting system fulfilled user expectations. Also, the project was completed on time and under budget. Proper implementation of the functional design approach should yield improvements in time, cost, and technical performance when compared to other approaches. The John Hancock project is of interest here because it encountered frequent (weekly) changes after the system was put into production and still managed to perform satisfactorily throughout. HIPO was originally conceived as a documentation tool (after the fact). Ironically, its greatest contribution may be as a functional design communication vehicle and project management estimating tool (e.g., amount of effort to produce the system).

Advantages of Using HIPO

Throughout this chapter we have spoken of structured design as something good. Some research supporting this has been referenced earlier in the text (e.g., *New York Times* and Skylab projects). Also, there are numerous other personal testimonies touting one or more of the structured programming tenets. The two general design approaches at each end of the spectrum are illustrated in Figure 5.8. Possibly this dichotemy is too harsh. Many traditional programmers do write good code and design good systems. However, far too often major system

Figure 5.7: Example of iterative refinement

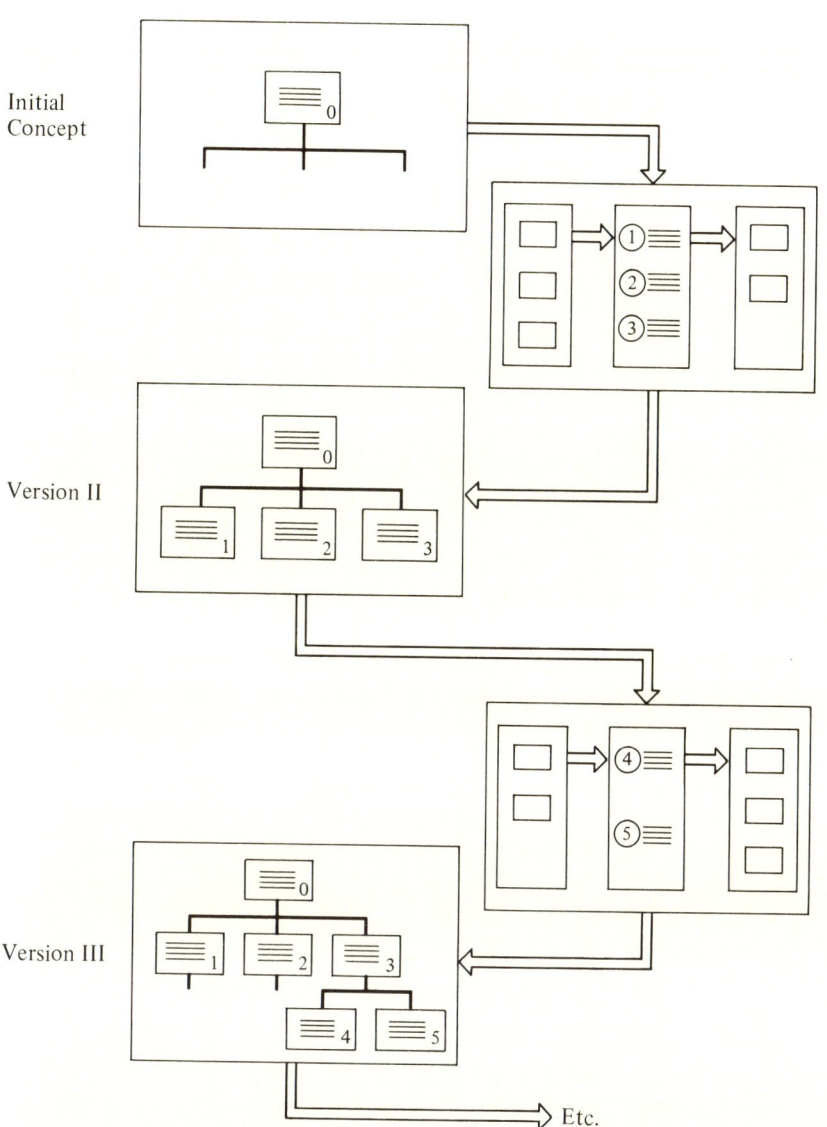

Initial Concept

Version II

Version III

Figure 5.8

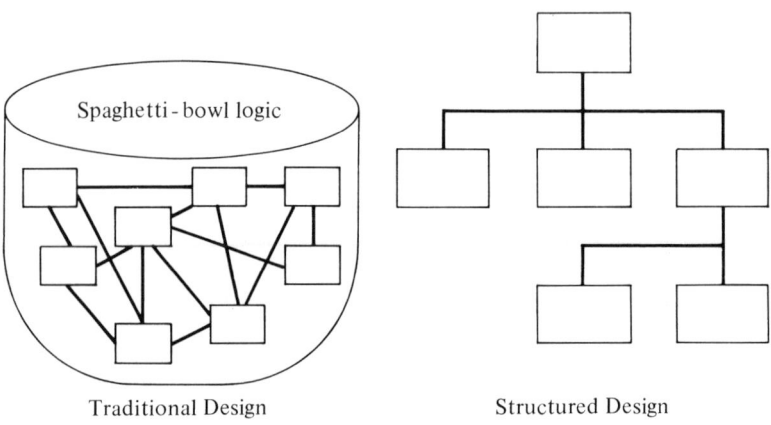

Traditional Design   Structured Design

segments cannot be modified by an organization because the original designer left, or the program interacts so much that one cannot tell what impact a change in one area has on the total program. Structured design rules try to minimize this and HIPO is one way to implement such an approach. Basically we are saying that a structured approach is better than an unstructured one. Table 5.1 summarizes some of the advantages from using a HIPO approach for design.

As stated before, *no* tool is a panacea. Many times "drowning" managers or programmers grasp for anything thrown at them and when it does not carry them swiftly out of the deep currents they become disenchanted and look for something else. HIPO must be used intelligently to be effective. Every program or system is not sufficiently complex to warrant this effort. However, for production systems of reasonable size the effort is justified and will pay great future dividends.

LOCAL IMPLEMENTATION OF HIPO

HIPO is not meant to be an academic subject with only one approved format and level of implementation. An individual or local installation should evaluate the tool in their unique environment and minimally decide the naming conventions and expansion of the HIPO diagrams.

Structured Design Techniques

Table 5.1. HIPO Advantages

1. Easier to read and maintain than a flowchart.
2. Enforces structured and top-down design philosophies by not being GOTO oriented.
3. Provides a common communication medium for both computer- and management-oriented personnel.
4. Generally supports the overall needs of functional design.
5. Helps to integrate management (functional) viewpoint and system structure early in the design.
6. Supports debugging, testing, and maintenance activities.
7. Supplies valuable information to the documentation process as a by-product of normal design activity.
8. Can be implemented to the degree desired (e.g., it is not an all or nothing tool).
9. There is no right way or wrong way to use HIPO, since it is more a set of conventions than an exact procedure.

Naming Conventions

Envision for a moment the fact that a HIPO box will eventually become a set of program logic. It is frequently useful to be able to make quick reference to a VTOC diagram or related detail diagram in conjunction with a set of code. This can show how various program pieces fit together, identify which logical segments call other logical segments, and so forth. Let us reuse Figure 5.6 to illustrate this point by assuming that each of the four logic processes referenced are accomplished by a PL/I procedure (or paragraph in COBOL). These four functions are summarized below with sample module names in parentheses:

1. Read transaction file and edit key fields (READ_TRANS)
2. Process edited file (PROCESS_FILE)
3. Sort data (SORT)
4. Tabulate results (TAB_RESULTS)

The names indicated for the four modules can be used to identify subsequent code. In addition to this the module numbering system becomes a convenient reference and source code sequencing key. Let

A Primer on Structured Program Design

Figure 5.9: Module naming convention example

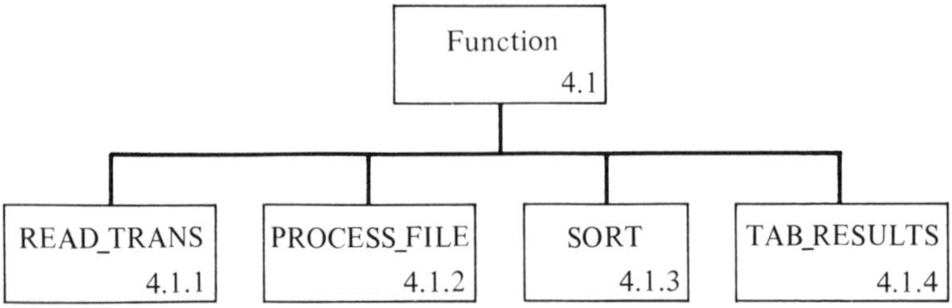

```
                    ┌─────────────┐
                    │  Function   │
                    │       4.1   │
                    └──────┬──────┘
        ┌────────────┬─────┴──────┬────────────┐
┌───────────┐ ┌────────────┐ ┌─────────┐ ┌─────────────┐
│READ_TRANS │ │PROCESS_FILE│ │  SORT   │ │ TAB_RESULTS │
│     4.1.1 │ │      4.1.2 │ │   4.1.3 │ │       4.1.4 │
└───────────┘ └────────────┘ └─────────┘ └─────────────┘
```

us show both of these ideas with the example referenced above.

A partial VTOC diagram which is logically equivalent to Figure 5.6 is shown in Figure 5.9. Module numbering can either be done sequentially (e.g., 10, 20, 30, etc.), or hierarchically as illustrated above. We will be consistent with the hierarchical type in this chapter but no preference is intended. As a matter of fact, the sequential approach is used in chapter 6 since most of the examples there are simple.

There are two major goals in this numbering scheme. First, to logically identify the module's function and, second, to provide some mechanism for finding the source code of a particular module or segment within a large listing. A module name thus consists of the following root parts:

sequence-name

Since PL/I and COBOL require a label to start with an alphabetic-type character, one slight modification is made to the above: all modules which follow the standard naming convention start with some selected letter such as M (module), P (procedure/paragraph), or S (segment). Using this convention, module 4.1.1 would have the following standardized name:

P4_1_1_READ_TRANS

Module source code is then stacked in ascending order of hierarchy which gives the listing a top-down orientation and makes selected modules easy to find. As we will see in chapter 6, some modules serve a utility (used several times) role in the overall system. These should be identified with an alternative key letter such as U (utility),

Structured Design Techniques

numbered sequentially and stacked in the back of a listing or kept completely separate.

Other guidelines can be used to improve retrievability of source code beyond the sequence logic illustrated here. For example, each module listing should be started on a new printer page and HIPO references within a module can be formatted on the right margin for improved visibility. This latter notion is not easy to demonstrate with the current example as is; however, if we assume that only the module referenced by the identifier 4.1 (see Figure 5.8) is a procedure and all other logic is imbedded within the procedure as more elementary logic segments, then the source references could be made using program comments as follows:

```
P4_1_FUNCTION: PROCEDURE;
       •
       •
       •                /* SEGMENT 4.1.1 */
                        /* SEGMENT 4.1.2 */
       (source code)    /* SEGMENT 4.1.3 */
       •                /* SEGMENT 4.1.4 */
       •
       •
END P4_1_FUNCTION;
```

Modules (procedures) such as this should generally occupy no more than one page.

There are numerous ways to help the reader understand what the module P4_1_FUNCTION does. The HIPO VTOC and detail-type diagrams should help. In addition to this some users believe that a brief description of a module's function should be included as a prologue comment in the source code. This can be abstractly demonstrated as

```
/*SORTS THE TRANSACTION FILE*/
/*AND PRODUCES REPORT X*/
P4_1_FUNCTION: PROCEDURE;
```

A second source of such information could be an alphabetical list of all modules with a brief statement of function (i.e., a module dictionary).

These and other such conventions should be implemented to improve the local design process, and all of the techniques discussed here have been helpful in various projects. In terms of programming

Figure 5.10: Overview diagram modified to show conditional logic

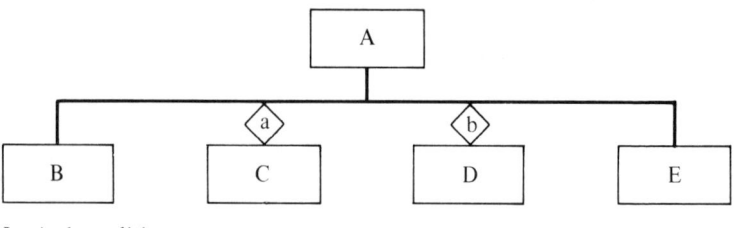

Logical conditions
   a — for transaction codes = 1
   b — for transaction codes = 2 or 3

procedures, each of these would seem to help reflect a good design, but will not force a bad one to get better.

Expansion of the HIPO Diagrams

As we discussed earlier, four basic types of code constructs are necessary to implement structured programming. These are:

1. Sequence
2. Condition (IFTHENELSE)
3. Iteration (DOWHILE, DOUNTIL)
4. Multibranch (case)

HIPO does little more than demonstrate sequence requirements, but with little modification could do a very nice job of indicating all of the structures required. Let us demonstrate how this can be added to the overview diagrams.

Condition. Figure 5.10 shows an abstract structure with conditional execution. In this example the logical structure specifies that once module A is entered module B will be executed (first). Module C will be executed when condition a is satisfied. Likewise module D will be executed for condition b. Finally, module E will be executed and control returned to module A.

IFTHENELSE (true-false) type logic can be represented as in Figure 5.11. In this way the higher level HIPO diagrams can migrate toward quite tangible code structures which have the familiar flowchart look without incurring its bad side effects.

Iteration. Repetitive execution of logic is the forte of a computer and must be shown to give proper perspective to a design. These two

Structured Design Techniques

Figure 5.11

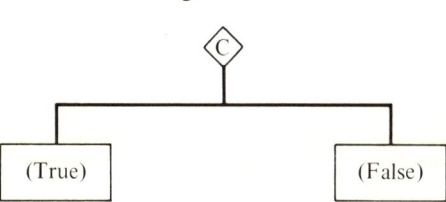

constructs are DOWHILE a specified condition is satisfied, and DOUNTIL a specified condition is satisfied. Figure 5.12 indicates that data is to be initialized, then a first record item is read. The iteration loop symbol indicates that TRANS1 or TRANS2 is to be executed depending upon logical condition a, results printed, and then a new data item read to restart the loop. This cycle continues as long as data is present. Finally, the data stream is exhausted and a program wrapup module completes the process. On a small scale this set of logic is quite realistic. Iterative loops such as the one illustrated here can be either of DOWHILE or DOUNTIL type.

Multibranch. When the number of related (nested) logical conditions exceeds three levels, a case-type construction can improve the code readability. For instance, assume that we had four different logical processes depending upon a transaction code (e.g., 1, 2, 3, or 4). A multibranch construct is appropriate for this, and Figure 5.13 schematically illustrates how the overview diagram can be modified.

Figure 5.12: Overview diagram modified to show iteration

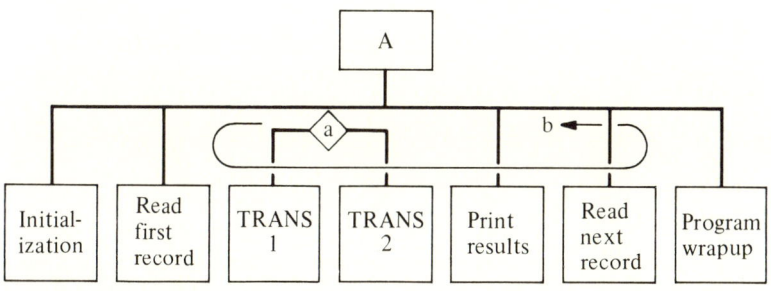

Logical conditions
   a — test transaction code for 1 or 2
   b — DOWHILE more data

Figure 5.13: Overview diagram modified for multibranch condition

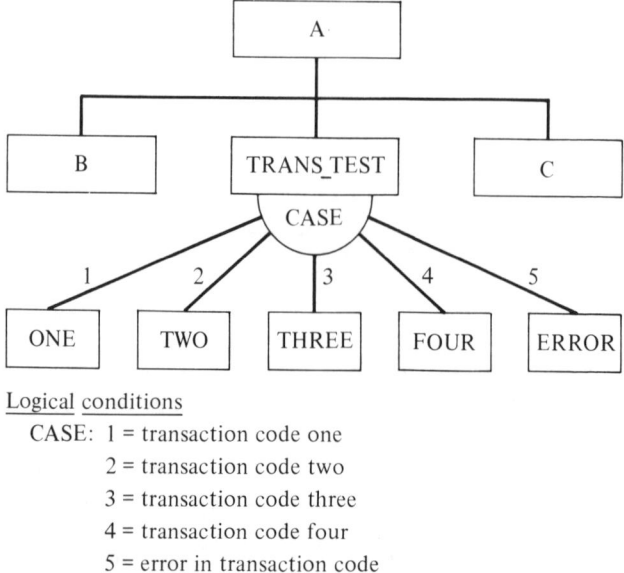

Logical conditions
   CASE: 1 = transaction code one
         2 = transaction code two
         3 = transaction code three
         4 = transaction code four
         5 = error in transaction code

In this final example only the selected module (ONE, TWO, etc.) will be executed depending upon the transaction code. Also, notice in this and most multibranch situations an ERROR module should be developed to handle faults in transaction codes.

Combinations. Of course, each of the logical types can occur simultaneously within the same structure. Collectively, they document the required flow of control necessary to implement the logic. Notice that the diagrams are implemented in a top-to-bottom and left-to-right fashion. Once properly constructed, it is quite easy to produce much higher quality programs.

SUMMARY

This chapter has attempted to address some of the more pragmatic issues of how to design a structured system which meets the theoretical requirements of structured programming. Much of this discussion revolved around the use of HIPO-type diagrams as vehicles to express a design in a top-down manner, followed by iterative refinement to evolve the lower level structures.

Structured Design Techniques

We have admitted that structured design still is an art, but less so than traditional approaches. Its key philosophy is to communicate the design so that others can comprehend its structure with minimal learning time. It also aids in functionally decomposing a system or program into smaller more comprehensible modules, which are in turn further decomposed until the logic is ready to be coded (or pseudo-coded). Chapter 6 is a companion discussion of design at the lowest level. Collectively, these descriptions of structured design and coding are meant to illustrate how to actually produce high-quality programs.

Review Questions

5.1. Define the acronym HIPO.

5.2 What are the three diagrams used in HIPO and what is the purpose of each?

5.3 Discuss the iterative refinement process.

5.4 What is the major purpose of HIPO in the system development process? Is this consistent with its original reason for development?

5.5. Why do flowcharts not readily support structured design?

5.6. Compare and contrast the vocabulary terms module, segment, module strength, and module coupling.

5.7. The two types of module numbering schemes discussed were hierarchical and sequential. Outline the perceived advantages and disadvantages of each.

5.8. Which portion of an IPO chart is developed first? Explain.

Notes and References

1. Mill, Harlan. *HIPO—A Design Aid and Documentation Technique.* White Plains, N.Y.: IBM Corporation, Form GC20-1851.
2. _____. *HIPO—A Design Aid and Documentation Technique.* White Plains, N.Y.: IBM Corporation, Form SR20-9413.
3. Jones, Martha. "HIPO for Developing Specifications." *Datamation,* March 1976, pp. 112–125.
4. Katzan, Harry. *Systems Design and Documentation: An Introduction to the HIPO Method.* New York: Van Nostrand Reinhold Co., 1976.
5. Omlor, Dennis. "Hierarchical Diagram Seen Better for Documentation."

Computerworld, March 28, 1977, p. 26.
6. Rubin, M.L. *Handbook of Data Processing Management, Volume I, Introduction to the System Life Cycle.* Princeton: Brandon/Systems Press, 1970.
7. Stay, J.F. *HIPO and Integrated Program Design. IBM Systems Journal*, No. 2, 1976, pp. 143–154.

6

PSEUDO-CODING

Pseudo-coding is a mixture of language-oriented control key words and English-like statements used to concretely describe an abstract design. It basically represents step three in the program design process as illustrated schematically in Figure 6.1.

As discussed here, there are four major identifiable goals associated with the pseudo-coding process. These are:

1. Aid in focusing attention on appropriate levels of design detail without becoming overwhelmed in minor low-level logic issues.
2. Provide a process which is very amenable to the creation of highly structured programs.
3. Replace difficult to produce and read flowcharts with English-like logic statements.
4. Provide a natural transition from high levels of logic abstraction into detailed code.
5. Facilitate program logic documentation process.

The goal of this chapter is to outline the pseudo-coding process, describe a semiformal set of syntax which aids in its implementation, and, finally, show how a mechanical formatting program can be used to improve the quality of documentation. Taken as an aggregate, the material developed here constitutes a structured design language.

Figure 6.1: Role of pseudo-code in program design

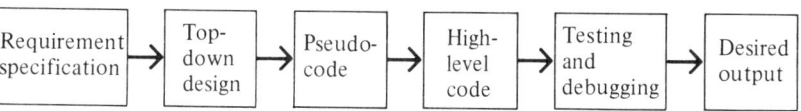

A Primer on Structured Program Design

SYNTAX STRUCTURE OF PSEUDO-CODE

The structure of pseudo-code is similar to a high-level language in that it contains key words for specifying the following:

1. Input/output
2. Conditional logic flow control
3. Iterative looping
4. Variable specification

On the other hand, a pseudo-code statement differs from a high-level language instruction by the following (note italicized words):

1. Pseudo-code syntax rules are more *flexible*.
2. *Levels of abstraction* can be varied to fit the particular state of a problem.
3. Pseudo-code *evolves* from high levels of design abstraction toward more rigorous code structure as logic segments become better defined.

The process of top-down design was discussed in chapter 3. From that discussion let us use an abstract design tree structure diagram to introduce pseudo-coding. Figure 6.2 contains a tree structure and associated pseudo-code. Notice in the introductory example that the design logic segments are developed from left to right. Also, since the design was constrained to express only structured constructs, the pseudo-code has no reason for excessive branching. Specifically, control logic structures are limited to the structured programming constructs:

1. Sequence—Use a separate line for each logical element and align each statement vertically until a change in logic flow occurs.
2. IFTHENELSE—Use the key words IF, THEN, and ELSE in pseudo-code format to express conditional operations. Provide a statement that ends the IF, such as ENDIF. Vertically align the IF, ELSE, and ENDIF for readability. Indent the statements contained in the IF and the ELSE to show the change in logic.
3. Iteration—Use the key words DO WHILE, DO UNTIL, DO FROM, TO, and BY as appropriate to define iterative control. Use an ENDDO to show block structure. Vertically align

Figure 6.2: Introductory pseudo-code example

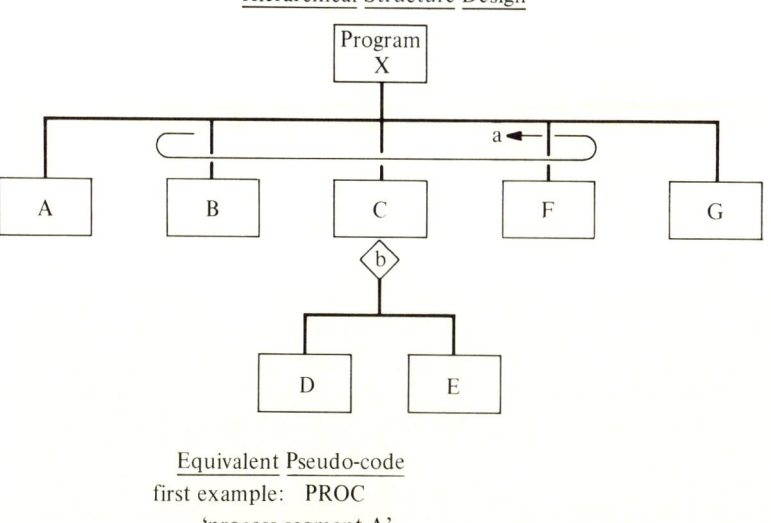

Equivalent Pseudo-code
first example: PROC
   'process segment A'
   DO WHILE 'condition a'
     'process segment B'
     'process segment C'
     IF 'condition B' THEN 'process segment D'
     ELSE 'process segment E'
     ENDIF
     'process segment F'
   ENDDO
   'process segment G'
ENDPROC   first-example

the DO and the ENDDO to show the change in logical structure.

4. CASE—Use the key words CASEENTRY and ENDCASE to indicate the start and termination, respectively, of a CASE structure. Vertically align the CASEENTRY and ENDCASE statements for readability.

Indentation of the control structures described in items 2, 3, and 4 above is often used to improve readability of the source code and show logic groupings. Tangible logic flow is expressed by the key words IF, DO, and CASE. Logic requirements not yet fully developed are represented by lower-case plain-language instruction; for example, if a record needs to be updated but the designer has not yet decided how to accomplish it, the designer simply indicates

'update master record'

as the appropriate point. Early stages of pseudo-coding emphasize *what* needs to be done and essentially *where* in the overall logic structure. Later efforts expand the early logic by defining variable names, correcting logic flaws, and elaborating necessary lower level code structures. In this manner, the pseudo-coding process is evolutionary starting with an abstract design structure and eventually stopping with a set of syntax which closely approximates a high-level language. The overall process proceeds in a top-down manner.

Saying that there is a typical set of pseudo-code for most programs is essentially the same as saying that all programs have a common flowchart. However, in many applications there is a surprisingly common logic pattern. As a simplified example of this, the sequence shown below assumes that multiple records are read and processed until all data is handled. When processing is complete, summaries are produced and the program is terminated. Thus, a typical pseudo-code sequence might be as follows:

'variable specification and initialization'
'open files'
'read first record'

DO WHILE 'more data'
   'process data'
   'read next record'
ENDDO
'wrap up processing'
'close files'

In the next section we will explore ways to more formally express pseudo-logic.

FORMALIZING THE SYNTAX

Now that a general philosophy of pseudo-coding has been established, a more formal, rigorous syntax can be defined. This syntax will be

Pseudo-Coding

Figure 6.3:   Design example (modified)

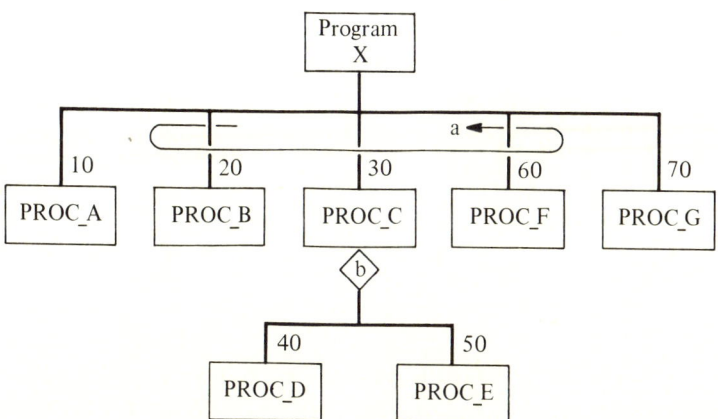

called *structured design language* and referred to as SDL. SDL is designed to fit the standard code blocks of structured programming theory and, additionally, it provides a flexible means of expressing logical functions until such time as the program is physically coded into a programming language. Figure 6.2 has already shown a simple example of SDL. The goal now is to illustrate a more complete set of key words and syntax.

SDL views program design as a collection of one or more procedures with a single entry and a single exit. Within the procedure there can be logical segments, variable specification, or other procedures. SDL primarily addresses the segment level of detail, although it is used to implement the hierarchical design as shown in Figure 6.2. To assist in describing this process and syntax, let us return to the design previously described, however this time assume that each segment is a subroutine and implemented by the key word CALL. Figure 6.3 shows this modified design. Notice in the modified design that each segment (PROC) has been assigned a reference number. Figure 6.4 contains an SDL implementation of this high-level design. Except for minor syntax this design is ready for translation into PL/I or another language.

Two major additions are illustrated in Figure 6.4. Upper-case letters are used for non-key words implying that a formal decision has been made regarding a procedure, file, or variable name. Lower-case identifiers signify that this has not been decided. Second, identification

Figure 6.4. SDL for Modified Design Example

```
DESIGN__EXAMPLE:     PROC
    CALL PROC__A                /* SEGMENT 10 */
    DO WHILE 'condition a'
        CALL PROC__B            /* SEGMENT 20 */
        CALL PROC__C            /* SEGMENT 30-50 */
        CALL PROC__F            /* SEGMENT 60 */
    ENDDO
    CALL PROC__C                /* SEGMENT 70 */

ENDPROC DESIGN__EXAMPLE
```

of segments is contained in the right-hand section of the SDL for reference back to the design and eventually forward into high-level code. These two conventions are very important and greatly enhance the evolutionary aspects of SDL. Basically, any instruction (statement) containing only upper-case items is ready to translate into formal code. Those that are not so designated need further refinement. Of course, the design example problem is not solved until each subroutine is completed. To accomplish this the decomposition process is repeated for each PROC.

The syntax for SDL is divided into four major categories: logic delineation, input/output, control structures, and variable specifications. Table 6.1 summarizes the essential key words in each of these areas. In addition to these key words, the following lexical symbols and conventions are used:

1. Lower case—signifies unexpanded items which do not fit high-level language constraints. These are designer-defined items.
2. Upper case—various SDL key words illustrating major logical items, or formally defined items such as variable or file names (i.e., indicate formal code structures).
3. Brackets [ ]—items are enclosed to show optional entries.
4. Braces { }—indicate a choice can or should be made between items enclosed.
5. /*comments*/—textual notes or segment delineation can be indicated by a /* at the beginning and a */ for termination. Thus, sample comments would be coded as

/* SEGMENT 10 */
/* BEGIN DO WHILE CYCLE */

The following sections will demonstrate these notions further.

Logic Delineation

The basic building blocks of SDL are procedures (PROC). Within this structure, variable specifications (DATA), input/output, and control structures as well as other procedures can be nested. General syntax is

>procedure name: PROC [segment type]
>•
>•
>•
>ENDPROC [procedure name]

where (a) procedure name is a descriptive title which identifies the segment, and (b) segment type optionally specifies the type of PROC. This is usually MAIN, INTERNAL, or EXTERNAL. More commonly the DATA option (see Table 6.1) is used internal to this procedure to identify variables.

Input/Output

Four different I/O options are identified in SDL. These are:

1. Sequential
2. Direct
3. Data base
4. Based

General syntax for each of these is shown below.

Sequential I/O. This option is the normal sequential access of file-oriented data. The general syntax is:

$$\begin{Bmatrix} \text{GET} \\ \text{READ} \end{Bmatrix} \text{FILE (filename)} \begin{Bmatrix} \text{(variable list)} \\ \text{'literal logic statement'} \end{Bmatrix}$$

$$\begin{Bmatrix} \text{PUT} \\ \text{WRITE} \end{Bmatrix} \text{FILE(filename)} \begin{Bmatrix} \text{(variable list)} \\ \text{'literal logic statement'} \end{Bmatrix}$$

Table 6.1. Structured Design Language Key Words

Operational Function	Key Word	Explanation
Logic delineation	PROCEDURE / PROC	Starts a logical segment
	ENDPROCEDURE / ENDPROC	Terminates a logical segment
		Acceptable abbreviation
	BEGIN	Starts a logical block within a segment
	ENDBEGIN	Terminates a begin block
Type of segment	MAIN	Indicates the main procedure
	INTERNAL	Indicates a local procedure
	EXTERNAL	Indicates a subroutine and requires argument passing
	DATA	Indicates a definition block describing variables, characteristics, and storage requirements
	ENDDATA	Terminates definition block
	COND	Indicates exceptional handling instructions
	ENDCOND	Terminates COND block
Input/output	GET / READ	Input instruction
	PUT / WRITE	Output instructions

	KEY	Implies direct access by a key variable
	FILE	Indicates logical file reference
	BASED	Pointer variable reference
Control structures	IF THEN	Conditional branch
	IFTHENELSE	Conditional branch (two ways)
	ENDIF	Terminates IF structure
	DO WHILE	Repetitive block
	DO UNTIL	Repetitive block that will be executed at least once
	DO FROM TO BY	Incremental block
	ENDDO	Terminates a DO block
	CASEENTRY	Starts a multiple branch
	ENDCASE	Terminates a CASE structure
	CALL	Invocation of a subroutine either INTERNAL or EXTERNAL
Data definitions	{CHARACTER, CHAR}	Indicates variable name to be alphanumeric characters
	NUMERIC	Indicates variable is numeric
	BOOLEAN	Indicates variable is of the true/false type
	FILE	Indicates logical file reference

Some selected examples of these are:

> GET 'next record'
> READ 'tape file'
> GET FILE(CARDIN) (X, Y, Z)
> READ FILE ('transaction') 'transaction record'
> PUT 'summary results'
> WRITE FILE ('output') 'summary'

The rule is to specify requirements to the level of detail known at that point in time. If the logical file name is defined, use capital letters to show it. If variable names are specified then they should be used. Otherwise code a "soft" requirement as indicated in the first and second examples above.

Direct Access. When direct access is needed, then add reference to this using the key word KEY. This is added to the end of a sequential statement as shown below:

> GET 'next record" KEY 'employee id'
> WRITE FILE (OUTPUT) 'new record' KEY SS__NUMBER

Data Base. Whenever a data base software package is being used, independence of data from program logic needs to be recognized. One way to do this is to write more abstract I/O instructions. One general form of such an expression could be

> c:=MEMBER (set__name) .TE.
> (D = 'value' & COUNT = 10 | type = 'ab')

which is read "let c be member of the data set defined as set__name. Within this set there exists (TE) a relation such that the contents of D are equal to 'value' and COUNT is greater than or equal to ten, or type of 'ab'." Due to the diversity of notation in various data base management systems, other syntax may have to be developed to fit local needs.

Based. Pointer-oriented I/O is generally restricted to a PL/I coding environment. In this mode there are a wide variety of options, however it is recommended that the requirement be specified simply as:

> GET   (variable)   BASED (pointer variable)
>   'variable'

Examples are:

>   GET (RECORD_A) BASED (PTRI)
>   GET 'transaction a' BASED ('pointer a')

Control Structures

Control logic within any SDL procedure is expressed through the key words CALL, IF, DO WHILE, DO UNTIL, DO, or CASEENTRY. When the key word CALL is used, it implies that control is passed to the procedure indicated and, upon its completion, is returned to the point from which it was called. Logic within a PROCEDURE block is expressed in terms of the control structure options outlined in Table 6.1 plus the sequential statement. A second control structure option is the IFTHENELSE which is constructed as follows:

>   IF 'condition is true' THEN DO
>   /*TRUE LOGIC BLOCK STATEMENTS*/
>   ENDDO
>   ELSE DO
>   /*FALSE LOGIC BLOCK STATEMENTS*/
>   ENDDO
>   ENDIF

The ELSE block may be omitted to form a simple IF THEN structure. Iterative looping within a set of logic is controlled by a DO WHILE block which is constructed by the following syntax

>   DO WHILE 'expression is true'
>   /*LOOP TO BE ITERATED*/
>   ENDDO

where the statements contained within the DO block may be combinations of SDL expressions.

Alternatively, an iterative block can be specified to cycle until a particular condition is encountered. In this case the logical structure is shown below:

>   DO UNTIL 'condition is false'
>   /*LOOP TO BE ITERATED*/
>   ENDDO

A third form of iterative block specification is the traditional incremental DO specification

```
DO 'control variable' FROM VALUE1 TO VALUE2 BY VALUE3
/*INCREMENTAL BLOCK*/
ENDDO
```

where VALUE1, VALUE2, and VALUE3 are the initial, ending, and incremental counter values, respectively.

The final SDL control construct is the CASE structure which involves multiple branching locations, entry to which is specified by variable 1. The SDL syntax for the CASE is

```
CASEENTRY variable1
    CASE (value1):
        •
        •
        statements(s)
        •
        •
    CASE(value2)
        •
        •
        statements(s)
        •
        •
    CASE(value3):
        •
        •
        statements(s)
        •
        •
ENDCASE
```

where (a) the value of "variable1" determines which case segment to execute and (b) the method used to branch from the case entry is not shown in SDL, but determined by the specific programming language used.

VARIABLE SPECIFICATIONS

Individual data names and their characteristics are eventually described in a logic segment with the DATA key word in conjunction with one of the data definition types. In addition to scalar variables, SDL allows the defining of related variables in a hierarchy form similar to the PL/I

Pseudo-Coding

structure or COBOL record, as well as array (table), boolean (bit), and alphanumeric forms. The data block is used to specify variable specifications, and the various key word options are summarized in Table 6.2. Only one potentially confusing reminder needs to be made regarding this portion of the pseudo-code. The assumption is made here that all variables are being formally defined and are therefore shown as capitals. Were this not the case, they should be shown as lower-case to signal that final specifications have not been made. Production of the data block typically begins after the broad flow of control logic is developed and concurrently with the lower level logic segment development.

General syntax for data block variable specification is

DATA
[(level)] variable name   type[(subscripts)]   [INITIAL (value)]
ENDDATA

where

a. level: is a logical identification of a heterogeneous aggregate of data such as COBOL record or PL/I structure.
b. type: is a key word denoting one of two categories used to define a data element (e.g., FILE, NUMERIC, or other data type from Table 6.1).
c. subscripts: represent multiple allocations of the same variable, as in an array.
d. value: is the initial allocation of the variable.
e. Bracketed items are optional.

For example, the specification of a two-element structure with ten arrayed items would be specified by

    1 INPUT_ITEM ARRAY (10)
    2 ITEM_A CHAR (10)
    2 ITEM_B NUMERIC

A sample DATA block is as follows:

```
        /*START OF DATA BLOCK*/
DATA
        /*SCALAR WITH INITIAL VALUE*/
    XVALUE NUMERIC INITIAL(10)
    VALUES(10) NUMERIC /*ARRAY*/
        /*STRUCTURE*/
```

Table 6.2. Data Block Key Words

Item Description	Type Key Word	Example
Homogeneous aggregation of variables	ARRAY (SUBSCRIPT)	ARRAY(10, 10)
Binary scalar	BIT (length)	FLAG BIT(1)
Generic built-in function	BUILTIN	SORT BUILTIN
String variable	CHAR (LENGTH)	NAME CHAR(20)
External entry point	ENTRY	MOD_A ENTRY
User-defined function	FUNCTION	FUNCTION INVERT
Statement label	LABEL	LAB_26 LABEL
Fixed or float scalar variable	NUMERIC	X NUMERIC
Establish initial value	INITIAL (value)	X NUMERIC INIT(10)
Used to access based variables	POINTER	P POINTER
Based variable	BASED (variable)	RECORD BASED(P)
Logical file	FILE	CARDIN FILE
System-defined routine	SYSTEM	SORT SYSTEM

```
1 CARDIMAGE
  2 ITEMA NUMERIC
    /*CHARACTER ELEMENT*/
  2 ITEMB CHAR(30)
  2 ITEMC NUMERIC
    /*CHARACTER STRING - LENGTH UNDEFINED*/
STRING CHAR
    /*CHARACTER STRING - LENGTH DEFINED*/
NAME CHAR(20)
ENDDATA
```

Throughout the pseudo-coding process the use of upper-case letters implies that a decision has been made as to variable names. For example, the expression

> DO WHILE 'more data'

implies that the method to define whether there is more data has not yet been decided. Conversely, the expression

> DO WHILE (MOREDATA = 1)

suggests that the variable MOREDATA is to be used to indicate whether more data exists to be processed. This variable will be set to a value of one while data remains to be processed. This brief example illustrates how pseudo-code evolves toward "hard" code. In this fashion SDL can serve as logical documentation for the program being developed.

FORMATTING CONSIDERATIONS

One final mechanical portion of SDL is its use of pseudo-code indentation to highlight code structures such as IF blocks and nested DO constructs. As an example of this, the set of pseudo-code shown in Figure 6.5 represents syntactically valid SDL. Certainly the collection of statements shown in this example may well describe necessary program logic, however it does little to highlight the natural logic structure. Figure 6.6 shows what indentation and formatting might do to enhance the same set of pseudo-code.

Two major benefits result from the modified format illustrated in Figure 6.6:

1. Logic segments are much easier to follow due to indentation.
2. Segment identification is made more valuable by using the

Figure 6.5:   Unformatted Pseudo-Code

```
ABLE: PROC
DATA
X NUMERIC
Y CHARACTER
ENDDATA
/*SEGMENT 10*/
GET 'first record'
DO WHILE 'more data'
/*SEGMENT 20*/
'perform calculations'
/*SEGMENT 30*/
PUT 'results'
/*SEGMENT 40*/
GET 'next record'
ENDDO
/*SEGMENT 50*/
'perform summary calculations
END A
```

right-hand columns and appropriate reference identification to the original design structure (e.g., /*SEGMENT 20*/).

CONDITION HANDLING

Most languages now give some measure of control over exceptional conditions. In some languages the only control option is where to branch when an input command aborts because the referenced file is empty. However, languages such as PL/I have extensive abnormal (error) handling facilities ranging from what to do when an end of page is reached on a printer to zero divide, conversion, transmission, overflows, and a host of other situations. SDL recognizes this class of logic development by introducing the key word COND which stands for condition. A logic segment designed to handle some specified abnormal condition can be constructed as follows:

```
COND condition name BEGIN
   /*ERROR HANDLING STATEMENTS*/
ENDCOND
```

Pseudo-Coding

Figure 6.6: Formatted Pseudo-Code

```
ABLE: PROC
  DATA
    X NUMERIC
    Y CHARACTER
  ENDDATA
  GET 'first record' /*SEGMENT 10*/
  DO WHILE 'more data'
    'perform calculations' /*SEGMENT 20*/
    PUT 'results' /*SEGMENT 30*/
    GET 'next record' /*SEGMENT 40*/
  ENDDO
  'perform summary calculations' /*SEGMENT 50*/
END ABLE
```

Use of the BEGIN key word signifies that multiple statements follow. If a single statement will suffice, the BEGIN can be omitted. A sample SDL statement to intercept a conversion error could be specified as:

>   COND CONVERSION
>   CALL RESTART
>   ENDCOND

A less rigorous example is considered to be executable, although it will

>   COND ENDFILE ('input file') BEGIN
>   'set end of file flag'
>   'set condition b flag'
>   ENDCOND

This type of statement is considered to be executable, although it will typically be set only once at the beginning of a program. It should not be imbedded in the logic definition unless required by the design. Obviously, the particular mechanics for this type of statement depends on language used, compiler, and local implementation.

USE OF COMMENTS

Comments have been used liberally in this discussion to illustrate various aspects of SDL, but there is disagreement as to how frequently

Figure 6.7. Comment Formatting Example

/*H PROCEDURE HEADING COMMENT*/
(top of each page and left justified, column one)

COM_ES: PROC   /*|SEGMENT 10*/
    •
    •
    •
LOOPA:

    •
  DO WHILE (EOF = 1)
    •
    •
    •
/*2 SORT SEGMENT*/
(justified to left margin, column one)
  XRAY = ALPHA + BETA

    /*3 EXPLANATORY COMMENT*/
    (imbedded within source margins)

---

they should be used in actual practice. Some feel that good code structures and appropriate variable names are the best form of internal logic definition. These conservatives feel that *no* comments are best. Liberal comments are used by others who feel that computer-oriented syntax can never be completely clear. This is analogous to the question "how far is it from here to there?" An obvious response is "it depends on how you go." So it is with comments. For purpose of this discussion four different types of comments are identified along with their location within the listing:

  H. Procedure (PROC) labeling (top of each page and left justified)
  1. Module or segment definition (right side of listing)
  2. SDL-oriented structure such as end of a particular segment (left statement margin)
  3. Other explanatory reader notes (indented or imbedded in the code)

We feel that these are useful in the order presented, that is, procedure and module definition comments should be mandatory, logic structure notes are generally helpful if not overdone, and explanatory notes should be used only in special situations. Also note that each type of comment is assigned to specific format locations within both the pseudo-code and subsequent source code listing. Using this format it is possible to scan the right margin to find particular logic segments easily by code reference to the original design structure. Once this is done major code segments are identified with type 2 comments on the left statement margin. Explanatory comments, type 3, would be kept off reading margins thereby making them visible only to the reader particularly interested in that logic segment. These notions are illustrated by the skeleton pseudo-code shown in Figure 6.7.

The SDL mechanical formatting program referenced earlier uses a coding convention as shown in this figure to handle comments. For example, the third character in the comment string contains the format logic. Selection of these rules has been somewhat arbitrary and could be changed to fit local needs. Most likely some disagreement would occur over using the left margin for comments since this is often saved solely for labels. One compromise to this is to indent the type 2 comment, thereby removing it from the margin. Our interest is more in standardization than specific choices of location. As currently written the format program assumes that any undesignated comment is of type 3.

PHILOSOPHY OF IMPLEMENTATION

Much of this chapter has been dedicated to showing specific syntax of a formal version of pseudo-code called SDL. It is hoped at this point that the reader has a reasonable idea of how to produce the various aspects of this design approach. Previous sections have addressed the following design areas:

1. Logic delineation
2. Input/output
3. Control structures
4. Variable specifications
5. Formatting considerations
6. Condition handling
7. Use of comments

This was done because of a firm belief that good program design and documentation practice must consider these essential issues, plus

A Primer on Structured Program Design

Table 6.3. Pseudo-Coding Guidelines
(Summary of Key Steps)

1. Create a tree structure design which logically exemplifies the design requirement.
2. Develop input, process, and output specifications for that part of the design which is to be implemented. (Ideally, this should be accomplished for the entire design, but can be fragmented if done with care.)
3. Develop broad flow of control logic first (see Figure 6.2).
4. Proceed in a top-down manner to develop each logic segment by assuming one function in and one function out (e.g., IF 'record type is A' THEN CALL 'update segment').
5. As logic development evolves, specific variable names chosen should reflect their meaning or function.
6. Major block segments should be terminated using the key words ENDPROC, ENDIF, ENDDO, ENDCOND, or ENDCASE.
7. Make logic constraints general using literal strings to describe requirements (e.g., 'compute average salary'). Refinements occur in an evolutionary manner. Don't get hung up on picky details too early in the design.
8. Use simple logic structures even at the expense of execution efficiency. Enhancements can be made during testing if improved efficiencies are needed. Normally 10% of the code consumes 90% of the time.
9. Comments are recommended to show design structure and complex code areas (refer to previous notes on this).
10. Coding activities can begin when *all* pseudo-code within a particular PROC is in upper case. Presence of lower-case code implies that further expansion is required and coding should not begin yet.

others such as individual style. Unfortunately, the attempt to provide some level of concreteness to the pseudo-code syntax may leave the impression that SDL, or any pseudo-code, is rigorous like a high-level language. This is *not* the case!

To be used effectively, local implementation rules for SDL must be allowed flexibility and should be modified as needed. The single most important reason for using this approach is to *drive* a design from the abstract toward tangible code structures in a particular language. This generally means that at some stage the designer is either going to have to choose a language or have one chosen for him/her. Once this occurs the design must be constrained by this language's capabilities. For example, one coding in FORTRAN would not specify hierarchical

Pseudo-Coding

structures, or a COBOL programmer would not implement pointer variables in his/her design. Bluntly stated, a particular language forces the program designer to lower his/her level of mentality to fit the selected language's capability.

Development Guidelines

The individual pseudo-code process has been viewed discretely to this point. Let us now attempt to establish a general set of guidelines upon which the SDL process is meant to operate. These are summarized in Table 6.3.

Various portions of this discussion and previous chapters have elaborated on the steps outlined in Table 6.3. A little practice with the pseudo-coding approach is usually sufficient to show its superiority over the traditional flowchart. Many programmers have been doing this informally over the past several years, then throwing it away after the program is coded. The proposition here simply says that this type of documentation is worth saving whether it be to help the designer during debugging, to share logic design with others, or subsequently to help the maintenance programmer. Remember that the average production program is handled by seven generations of programmers. It should *not* be written with a philosophy that only the original designer need understand it.

Developing a Logic Segment

The most important initial consideration of logic segment development is that a designer fully understand the external conditions upon which it is to operate. This means that the logical algorithm must be well defined (often easier said than done). The HIPO approach previously outlined has given us procedural tools to sketch out the key inputs, processes, and outputs related to a particular design. If this is done properly the logical development activity is easier.

Whenever a logic segment is being designed the following items should be reviewed:

1. What data variability can be expected (ranges, erroneous values, etc.)?
2. How complex and well defined is the logical process?
3. If iteration is required, have the following items been considered:
    a. Control of the iterative process (e.g., what stops it)?
    b. How is the loop entered (e.g., first cycle)?

c. How is the loop exited (e.g., last cycle)?
4. What major logical conditions can be expected?
5. What unusual logical or processing situations might occur (e.g., null file, zero divide, effects of bad data, etc.)?
6. What boundary type conditions should be tested or handled?

Obviously, it is presumptuous to write six steps which will handle everything to be known about logic development. These are simply a sample of initial mental items which should be considered. New programmers tend to feel that the "programming fairy" would never send them anything bad or unusual, while older programmers view the design process as "I wonder what crazy things are going to happen this time." The point being made here is that proper design of a logic segment can at least isolate the severity of problems to be handled. All key assumptions should be explicitly recorded as part of the program documentation.

FORMAT AND VERIFIER PROGRAM

Throughout much of this discussion the issue of formatting has been discussed as being a helpful part of the readability and documentation process. Unfortunately, programmers do not want to spend a great deal of time indenting and formatting a pseudo-code listing according to some standardized rules. Indeed this is not a very creative exercise. Most organizations that have attempted to implement this process formally have found that a utility program to indent and format source code or pseudo-code is helpful. This section will briefly describe a utility program to process unformatted pseudo-code and produce formatted output. Additionally, the program accomplishes the following:

1. Analyzes SDL statements, indenting them for readability, noting key word omissions, and generally providing feedback on style (e.g., block structures and use of GOTO statements).
2. Provides statistical information such as
   a. number of statements processed
   b. number of procedures
   c. number of statements in each procedure
   d. list of all CALL statements
   e. list of GOTO statements
   f. number of code blocks

g. SDL error messages
h. warning messages regarding style

This program is designed to eliminate the need for a detailed flowchart. Whether the SDL format/verifier program is used, or a locally written simple indentation program, the purpose is to provide standardized pseudo-code formats for debugging and program documentation. A project librarian should be responsible for cataloging past and current versions of the design pseudo-code. A configuration version number can be used to keep track of the evolution of code structures for a particular logic segment.

Upon completion of the pseudo-code process the design is essentially ready to be implemented into a higher level language. The SDL format/verifier program allows completed PL/I source statements to be imbedded in the pseudo-code. As code development progresses the format program will provide proper structural indentation until the final high-level code is completely produced and debugging begins. At this point a compiler can be used to process the source statements. The design example presented below further describes the pseudo-code process and format program output.

A DESIGN EXERCISE

Suppose the GLR Corporation is trying to develop a program to update its charge account file. A systems (file) level flowchart depicting this process is shown in Figure 6.8. The master file contains the following data elements:

1. Customer name
2. Account number
3. Current balance

Transaction data consists of the following data elements:

1. Customer name
2. Account number
3. Payment/charge amount
4. Transaction code (1 = payment, 2 = charge purchase)

To simplify logic development, assume that both input files have previously been sorted into ascending account number order. Procedurally, it is necessary to read a master record, then match it with a transaction record. After a match is found the transaction should be

Figure 6.8

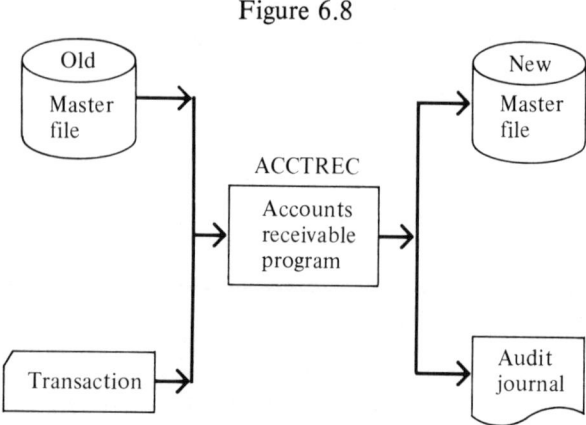

processed. Next, the new master file record is created and the logic cycle repeated. Two iterative loops are identified. First, all master records must be processed and it is assumed that the transaction account numbers are no higher than those contained on master records. Second, an internal loop is required for all transactions related to a single master record. For example, a customer might have numerous activities against his file in a single update cycle. After all master records are processed a final section of logic is needed to complete the program. Version I of this logic is represented by the hierarchical design structure shown in Figure 6.9.

Closer examination of the logic developed in Version I raises the following questions:

1. Is the loop control logic correct (e.g., loops a and b from Version I)? Does it accomplish the desired result for first record, iterative records, and last record?
2. What type of ENDFILE conditions should be anticipated for the master and transaction files?
3. Should transaction code errors be anticipated?

Each of these points needs to be researched, plus some additional thought given to file and variable names. Let us name the three files required by the design as follows:

OLDMAST—old master file

TRANS—transactions file

NEWMAST—new master file

Figure 6.9: Hierarchical structure—version I

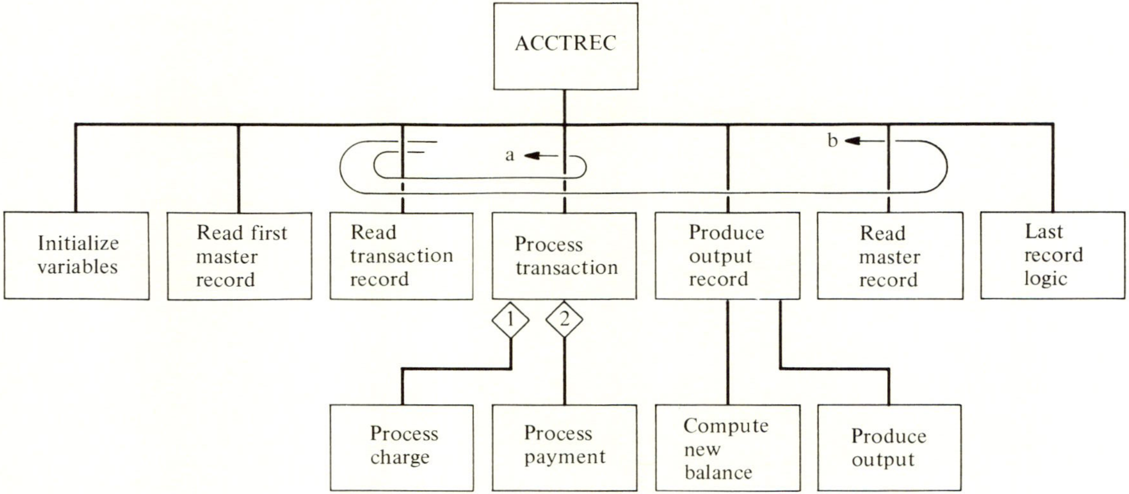

Logical Conditions

a – while more data remains on master file
b – while transaction account matches master file record
1 – transaction code = 1
2 – transaction code = 2

The printer will be used for audit output. Logic flags will be set for each input file to indicate ENDFILE status. These flags are OLDFLAG and TRANSFLAG, respectively. In similar fashion other needed variable names are selected and these will be added to the SDL DATA section in Version II.

For the next phase let us now resolve the logic issues itemized above as:

1. Logical conditions a and b will be assumed adequate as initially stated.
2. ENDFILE conditions can be expected on either file depending upon data values, although they will most likely occur on the transaction file first.
3. Transaction codes other than one or two should be handled as errors and printed on the audit journal only (e.g., not processed).

With these changes implemented, a Version II pseudo-code is presented in Figure 6.10.

Let's now carefully evaluate the specified flow of control outlined in Version II under various assumptions. Several logical questions arise and are summarized below:

1. Logic produced does not properly handle the situation where TRANS_LOOP is entered and accounts (master file and transaction file) do not match.
2. Location of initialization for various accumulators cannot be only at beginning. It must also be at appropriate account break points.
3. Current program logic could produce erroneous results if a keypunch error was made in creating the transaction file. For example, a transaction account number higher than the last master record would cause serious problems.

What other logic flaws can you find in this design?

Version III of the pseudo-code resolves logic issues one and two. It is arbitrarily decided to ignore account number errors on the transaction file in order to keep the example from growing in complexity. Figure 6.11 presents the revised logic.

Whether one uses a revised schematic top-down structure to produce the pseudo-code or manually evaluates the pseudo-code and from it revises the initial top-down design seems like a moot question.

Pseudo-Coding

Figure 6.10: Accounts Receivable Pseudo-Code—Version II

```
/*HVERSION II ACCOUNTS RECEIVABLE PSEUDO CODE*/
DATA
   ACCTREC:   PROC MAIN
   OLDMAST    FILE
   TRANS      FILE
   NEWMAST    FILE
   OLDFLAG    BINARY
   TRANSFLAG
   NAME       CHAR
   TRANS-CODE NUMERIC
   ACCT       NUMERIC
   OLD-BAL    NUMERIC
ENDDATA
   'initialize variables'
   'read master record'
MASTER_LOOP:
   DO WHILE 'OLDFLAG not raised'
   'read transaction record'
TRANS_LOOP:
      DO WHILE 'TRANSFLAG not raised & accounts match'
         IF TRANS_CODE=1 THEN
            'accumulate payments'
         ELSE IF TRANS_CODE=2 THEN
            'accumulate charge'
         ELSE
            'process error transaction code'
         ENDIF
         'read transaction record'
      ENDDO /* TRANS_LOOP */
      'compute new balance'
      'create new master file record'
      'write results to audit journal'
      'read master record'
ENDDO /* MASTER_LOOP */
'last record logic'
ENDPROC
```

The important point here is that the design at this stage properly reflects the required logic steps and flow of control is accurately identified (e.g., IF- and DO-type logic). Also, any logical assumptions

A Primer on Structured Program Design

Figure 6.11: Accounts Receivable Pseudo-Code—Version III

```
/*HVERSION III ACCOUNTS RECEIVABLE PSEUDO CODE*/
ACCTREC:   PROC MAIN
   DATA
      (same variables and files previously shown)
   ENDDATA
   'initialize variables'
   READ FILE (OLDMAST) 'master record'
   READ FILE (TRANS) 'transaction record'
MASTER_LOOP:
      DO WHILE OLDFLAG 'not raised'
         DO WHILE 'master account not equal trans account'
            WRITE FILE(NEWMAST) 'from old master record'
            READ FILE(OLDMAST) 'master record'
         ENDDO
TRANS_LOOP:
         DO WHILE 'master account=trans account'
         & TRANSFLAG 'not raised'
            (same code as previously shown in Version II)
         ENDDO /*TRANSLOOP */
         'compute new balance'
         WRITE FILE(NEWMAST) 'updated record'
         'write results to audit journal'
         'reinitialize accumulators'
         READ FILE(OLDMAST) 'master record'
      ENDDO /*MASTER LOOP */
      'end of file processing'
ENDPROC
```

which may affect processing results should be recorded in the user documentation. For example, we have assumed here that the transaction file is sorted and its account number ranges exceed those found in the master file. This limitation should be understood by the user. Many times a good sort/edit program can do much to isolate a program from undue variety in the data. For example, by using an edit program bad account numbers could be identified in the transaction file and removed before processing. Also, certain keypunch errors could be found and corrected before running the program.

It is left as a reader exercise to redraw the tree structure design which fits the pseudo-code developed in Version III (Figure 6.11). It would be an even better exercise to then translate the pseudo-code

Pseudo-Coding

into high-level code. There is still one very serious logic flaw remaining in the Version III solution. You should be able to discover this flow of control error in ten minutes or less.

SUMMARY

In this chapter we examined the role of English-like logic statements called pseudo-code in producing high-quality code from a hierarchical tree structure design schematic. A formal set of pseudo-code syntax called structured design language (SDL) was described to facilitate standardized description of data and logical relationships such as:

1. Logic delineation
2. Input/output
3. Control structure
4. Data definitions (files and variables)

Use of an SDL format/verifier utility program aids in producing high-quality documentation rather than an after the fact process. This utility formats both SDL and imbedded PL/I code, plus maintains high-quality source code until a compiler can be used to begin executing the code.

The design example presented has illustrated in some detail how to take a problem statement and evolve a structured set of code.

Problems

6.1. Produce a set of pseudo-code with accompanying top-down structure which will read an integer, test it for positive or negative, and print out appropriate results. Be sure to catalog any shortcomings in your design or data assumptions which are made.

6.2. A series of data regarding sales performance of salesmen is to be used for a management report. Key variables are employee number, name, department, and amount sold. Create a program design which will list this data alphabetically by department giving subtotals for each department and the organization.

6.3. Write a set of pseudo-code to scan text material punched on cards and identify the frequency of occurrence of alphabetic characters. At the end of the text produce an output summarizing the results.

6.4. The general form of a quadratic equation is ax^2+bx+c. Solution roots can be obtained by the formula

A Primer on Structured Program Design

$$x = \frac{-b \pm \sqrt{b^2 - 4ac}}{2a}$$

Read in the three parameters (a, b, and c) and compute the roots for each rational number. When the discriminant, $b^2 - 4ac$, is negative, the roots are complex and need not be computed. Write the pseudo-code to perform this computation for several sets of data. Also, proper headings should be provided.

6.5. Assume that the three sides of a triangle are defined by the variables a, b, and c. Develop a set of pseudo-code which will read three values for a hypothetical triangle. Analyze these values for the conditions:
   a. Valid triangle (e.g., one side not greater than the sum of the other two sides)
   b. Isoceles triangle
   c. Equilateral triangle

   Hint: Spend some time reviewing the various conditions and combinations which can occur before attempting to structure the logic.

CASE STUDY

A local clothing store puts all charges and payments to credit accounts on punch cards and from this wants to design a computerized charge account journal to calculate each customer's balance given various charge transaction data. Assume that all records are sorted such that there is a customer master record followed by a series of transaction records. At the end of each customer set is a dummy transaction record with "999" in the account number field.

The master record contains the following data:

Name (Columns 1–20)

Account number (Columns 25–34)

Beginning balance (Columns 40–49)

Transaction records have the following format:

Name (Columns 1–20)

Account number (Columns 25–34)

Transaction code (Column 40)

Amount (Columns 50–59)

A transaction code of 1 indicates payment on account, while a 2 indicates a new charge. For the sake of simplicity, assume that master

Pseudo-Coding

and transaction records are merged into a single sequential file. Further, assume that each master record has at least one transaction record.

The primary purpose of this example is to illustrate structural and top-down design procedures and not necessarily invent the best charge account monitoring program. Concepts illustrated here work well with large programs and the resultant product should not contain significant errors. When they do occur, location and repair is designed to be easy using the design documentation illustrated.

In order to test your understanding of this logic/documentation process, assume that the program description above has been given to you for incorporation. Create the necessary program logic and include the following items:

1. Print out the original balance under an appropriate heading.
2. Produce an appropriate heading.
3. Write each transaction entry under an appropriate column, then show the final totals as "monthly totals."
4. Test each transaction record to verify that it matches the master name and account number. If not, list the card contents and print an error message.
5. Test each transaction code and when invalid list the card contents and print out an error message.

Program Testing

Make up a set of test data with selected errors which you feel will validate the modified program's logic. List all the logical tests performed and state results. Try to make sure the program does not terminate until all data are examined.

Version I Design Structure and Pseudo-Code

Figure 6.12 illustrates the initial design structure and corresponding pseudo-code. Analysis of this design quickly identifies four logical problems:

1. Handling of the first record is not proper (e.g., the first record must be read outside the master loop).
2. A logical method of handling the end-of-file condition needs to be designed for the master loop.
3. Logical sequence for handling a new master record needs further work.

Figure 6.12a: Version I—Case Study Solution (initial hierarchical design)

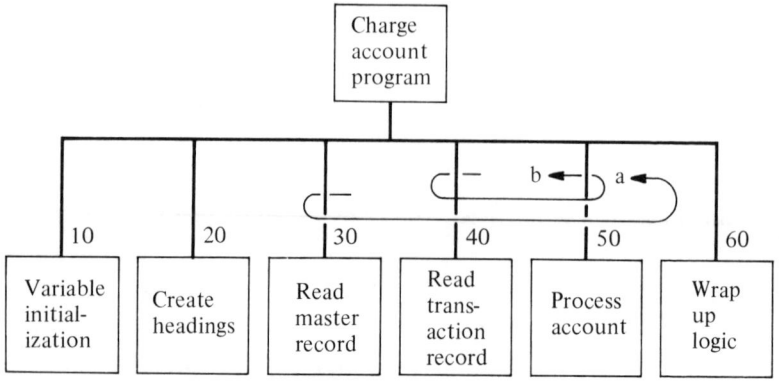

Logical Conditions

a — DO WHILE more records
b — DO WHILE account remains unchanged

(a)

4. Key variable names need to be defined.

From these identified deficiencies Version II of the pseudo-code can be produced. In the interest of space conservation a new design structure will not be presented, however it should be kept updated in actual practice (renumbering of segments is still necessary at this stage to fit changes). Figure 6.13 shows the modified design pseudo-code.

After the Version II logic design is completed a more detailed set of logic questions emerges. One group who looked at this design brainstormed the following logic questions for possible modifications:

1. What happens if the end-of-file condition occurs when trying to read transaction *a* card?
2. What happens if transaction code (TRANS) is punched as a 3?
3. What happens if the transaction terminator (e.g., 999) is mispunched?
4. Where would you modify logic to print out each individual transaction?

Figure 6.12b: Version I—Case Study Solution (pseudo-code)

```
/*H VERSION I PSEUDO CODE */

STMT   LEV  NEST        SOURCE PROGRAM TEXT

              #   /*H VERSION I PSEUDO CODE */
    2    1   #   CHARGE: PROC;
              #
    3    1   #     'VARIABLE INITIALIZATION'                  /*1 SEGMENT 10 */
              #
    4    1   #     'CREATE HEADINGS'                          /*1 SEGMENT 20 */
    5    1   #   MASTER_LOOP:
              #
    7    1   1   #   DO WHILE'MORE RECORDS'
    8    1   1   #     READ'MASTER RECORD'
    9    1   1   #   TRANS_LOOP:
   11    1   2   #     DO WHILE 'ACCOUNT NUMBER UNCHANGED'    /*1 SEGMENT 30*/
              #
   12    1   2   #       READ 'TRANSACTION RECORD'            /*1 SEGMENT 40*/
              #
   13    1   2   #       'PROCESS ACCOUNT'                    /*1 SEGMENT 50*/
   14    1   1   #     ENDDO TRANS_LOOP
   15    1   1   #   ENDDO MASTER_LOOP
              #
   16    1       #   'WRAP UP LOGIC'                          /*1 SEGMENT 60*/
   17            #   ENDPROC CHARGE
```

(b)

/*H VERSION II PSEUDO CODE */

Figure 6.13: Version II—Case Study Solution

```
STMT  LEV NEST       SOURCE PROGRAM TEXT

             #  /*H VERSION II PSEUDO CODE */
     19   1   #  CHARGE: PROC;
     20   1   #     DATA
             #        NAME                          CHAR
             #        TRANS                         NUMERIC
             #        ACCT                          NUMERIC
             #        BAL                           NUMERIC
             #        AMOUNT                        NUMERIC
             #        CHARGE                        NUMERIC
             #        PAYMENT                       NUMERIC
     21   1   #     ENDDATA
             #
     22   1   #     'VARIABLE INITIALIZATION'            /*1 SEGMENT 10 */
             #
     23   1   #     'CREATE HEADINGS'                    /*1 SEGMENT 20 */
             #                                           /*1 SEGMENT 30 */
```

```
24  1    #       'READ FIRST MASTER RECORD'
25  1    #  MASTER_LOOP:                                           /*1 SEGMENT 40*/
         #
27  1    #    DO WHILE'MORE RECORDS'
28  1    #      'READ TRANSACTION RECORD                           /*1 SEGMENT 50 */
29  1    #  TRANS_LOOP:                                            /*1 SEGMENT 60 */
         #
31  1 2  #      DO WHILE (ACCT¬=999)
         #
32  1 2  #        IF TRANS=1 THEN 'ACCUMULATE PAYMENTS'
33  1 2  #        ELSE 'ACCUMULATE CHARGES'
34  1 2  #        'READ TRANSACTION RECORD'                        /*1 SEGMENT 70*/
         #
35  1 2  #      ENDIF
36  1    #      ENDDO TRANS_LOOP                                   /*1 SEGMENT 80 */
         #
37  1    #      BAL=BAL+CHARGE - PAYMENT                           /*1 SEGMENT 90 */
         #
38  1    #      'WRITE OUTPUT RECORD'
39  1    #    ENDDO MASTER_LOOP                                    /*1 SEGMENT 100 */
         #
40  1    #    'WRITE EOJ TRAILER MESSAGE'
41       # ENDPROC CHARGE
```

Figure 6.14: Version III—Case Study Solution

```
/*H VERSION III PSEUDO CODE */

STMT LEV NEST       SOURCE PROGRAM TEXT

              #  /*H VERSION III PSEUDO CODE  */
              #  /* CHARGE ACCOUNT LOGIC */
   43   1    #  CHARGE: PROC;
   44   1    #     DATA
              #        NAME                              CHAR
              #        TRANS                             NUMERIC
              #        ACCT                              NUMERIC
              #        BAL                               NUMERIC
              #        AMOUNT                            NUMERIC
              #        CHARGE                            NUMERIC
              #        PAYMENT                           NUMERIC
   45   1    #     ENDDATA
              #     /* INITIALIZE VARIABLES */
   46   1    #     PAYMENT=0;
   47   1    #     CHARGE =0;
              #     /* HEADING */
```

```
48  1    #  'WRITE HEADING'
         #  /* FIRST CARD */
49  1    #  'READ FIRST MASTER CARD'
         #  /* MASTER CARD LOOP */
50  1    #  'READ TRANSACTION CARD'
         #  /* TRANSACTION CARD LOOP */
         #  TRANS:
51  1    #  DO WHILE (ACCT ¬=999)
52  1 1  #    IF TRANS = 1 THEN DO;
53  1 2  #      'ACCUMULATE PAYMENTS'
55  1 2  #    ENDDO                    /* TRUE BLOCK */
56  1 1  #
57  1 2  #    ELSE DO;
59  1 2  #      'ACCUMULATE CHARGES'
60  1 1  #    ENDDO          /* ELSE BLOCK */
61  1 1  #    ENDIF
62  1 1  #    'READ TRANSACTION RECORD'
63  1    #  ENDDO  TRANS       /* TRANSCATION LOOP */
         #  /* SUMMARY SECTION */
64  1    #  BAL=BAL+CHARGE -PAYMENT;
65  1    #  'WRITE OUTPUT RECORD'
66  1    #  'WRITE EOJ TRAILER'
67       #  ENDPROC CHARGE
```

119

A Primer on Structured Program Design

5. How could you modify program logic to print out a new heading on each new page?
6. Where should CHARGE and PAYMENT be initialized?
7. Can you think of any other potential logic flaws in this?

Several design conclusions need to be reached regarding each of these items and we need to comment further on each item. First, an end of file reached when trying to read a transaction card is obviously an error in codes or sequence. (For the sake of maintaining a simple example this will not be handled here.) Second, a coding error for TRANS will have quite serious effects on the logical results of this program. The final design version will show where such a test should be inserted, but this error is essentially ignored in Version III. Third, a coding error in the terminator card will affect the TRANS_LOOP by causing it to be entered one extra time. No logical testing is done to identify this problem. Fourth, printing of individual transactions would be handled within the TRANS_LOOP—after the process segments (SEGMENT 70 in Version II). Fifth, new headings can be created either by inserting line counter logic within the code or by using the PL/I ON ENDPAGE instruction. This will not be included in the design example. Finally, there appears to be some subtle problem regarding placement of zero initialization for the variables CHARGE and PAYMENT. Version III will simply initialize them at the beginning of execution. Examine each of these issues realizing that they represent either designer perceptions of what is expected in the data input stream (e.g., coding and sequence errors), or location of initialization for key variables.

Version III of the design corrects or assumes away the general problems discussed above. Figure 6.14 illustrates the design configuration for this stage of development.

At this point in the evolving design, attention becomes increasingly focused on a few logical issues requiring repair in the next iteration. In this case the problem of reinitializing various accumulators and balances remains to be fixed. It is also advisable to record a list of exceptional conditions which are not picked up by the current design. This should be reviewed at each step and published in the final documentation.

Let us assume that Version IV of this design is represented by the HIPO structure of Figure 6.15a and the corresponding PL/I program shown in Figure 6.15b. It is hoped that at this point the general

Figure 6.15a: Version IV—Case Study Solution (design structure)

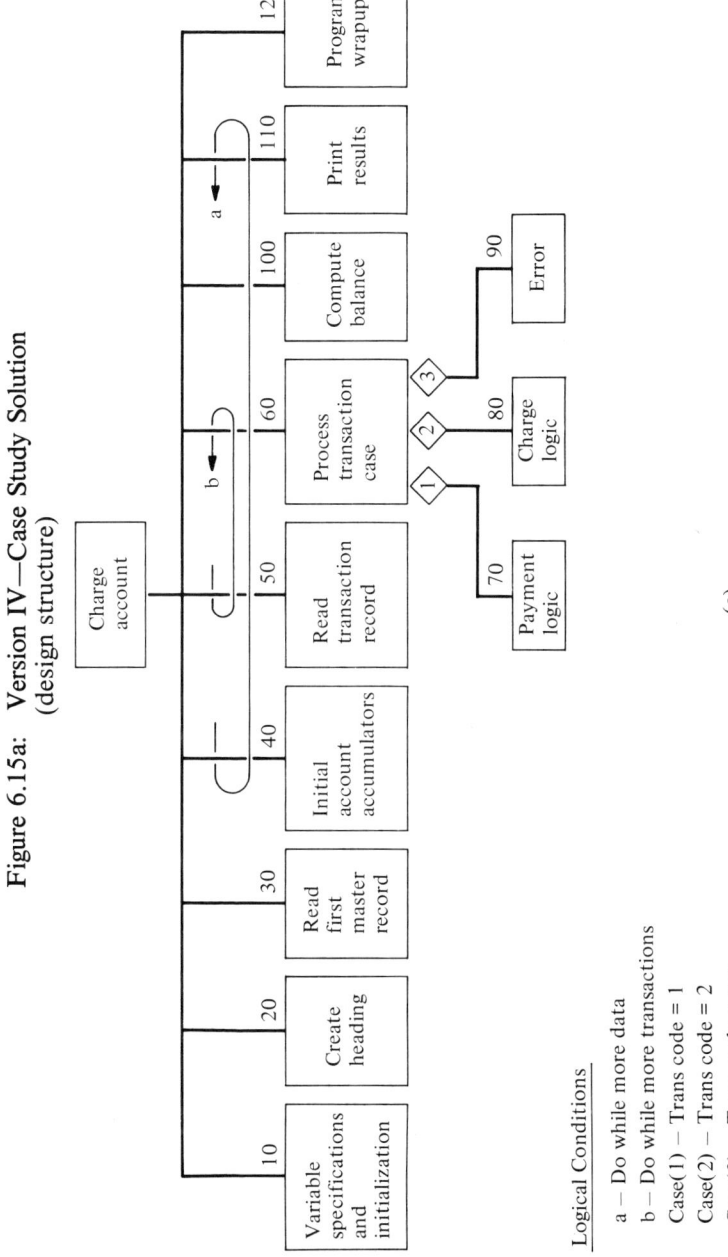

Logical Conditions

a — Do while more data
b — Do while more transactions
Case(1) — Trans code = 1
Case(2) — Trans code = 2
Case(3) — Trans code error

(a)

Figure 6.15b: Version IV—Case Study Solution (final PL/I code)

```
/*H VERSION IV DESIGN */

STMT  LEV NEST          SOURCE PROGRAM TEXT

                    #  /*H VERSION IV DESIGN */                                    /*1 SEGMENT 10 */
                    #
   1    1           #  CHARGE: PROC OPTIONS (MAIN);
   2    1           #    DECLARE
                            NAME                                   CHAR (20),
                            (BAL,CHARGE,PAYMENT,AMOUNT)            FIXED (10,2),
                            TRANS                                  FIXED BINARY,
                            END_TRANS                              CHAR(3) INIT('NO'),
                            ACCT                                   FIXED (10),
                            END_OF_FILE_FLAG                       BIT(1),
                            TRUE                                   BIT (1) INIT ('1'B),
                            FALSE                                  BIT (1) INIT ('0'B);
                    #    /* VARIABLE AND CONDITION INITIALIZATION */               /*1 SEGMENT 20 */
   3    1           #    END_OF_FILE_FLAG = FALSE;
                    #
   4    1           #    ON ENDFILE (SYSIN) END_OF_FILE_FLAG=TRUE;                 /*1 SEGMENT 30 */
                    #
   5    1           #    CALL HEADING;
   6    1           #    GET EDIT (NAME,ACCT,BAL)
   7    1           #        (COL(1),A(20),COL(25),F(10),COL(40),F(10,2));
   8    1           #  MASTER_LOOP:                                                /*1 SEGMENT 40 */
                    #
  10    1    1      #    DO WHILE (END_OF_FILE_FLAG=FALSE);
                    #    /*2 INITILIZATION STEP */
  12    1    1      #      PAYMENT = 0;
  13    1    1      #      CHARGE = 0;
  14    1    1      #  TRANS_LOOP:                                                 /*1 SEGMENT 50 */
                    #
```

```
16   1  2  #       DO WHILE (END_TRANS ¬= 'END');
17   1  2  #         GET EDIT (NAME,ACCT,TRANS,AMOUNT,END_TRANS)
18   1  2  #         (COL(1),A(20),COL(25),F(10),COL(40),F(1),COL(50),F(10,
19   1  2  #         2),COL(78),
20   1  2  #         A(3));                                    /*1  SEGMENTS  60-90   */

21   1  2  #         IF TRANS= 1 THEN
22   1  2  #           PAYMENT =PAYMENT+AMOUNT;
23   1  2  #         ELSE IF TRANS= 2 THEN                     /* CHARGE */
25   1  2  #           CHARGE =CHARGE + AMOUNT;
26   1  2  #           ELSE;            /* INSERT ERROR CODE HERE */
27   1  2  #           ENDIF
28   1  2  #         ENDIF
29   1  1  #       ENDDO TRANS_LOOP
           #  /*3 CACULATE NEW BALANCE   */                    /*1  SEGMENT 100    */
           #                                                   /*1  SEGMENT 110    */
30   1  1  #       BAL = BAL + CHARGE - PAYMENT;
31   1  1  #       PUT EDIT (NAME,ACCT,BAL,CHARGE,PAYMENT)
32   1  1  #       (COL(1),A(20),COL(25),F(10),COL(40),COL(40),3 F(12,2));
33   1  1  #       END_TRANS = 'NO';
34   1  1  #       GET EDIT (NAME,ACCT,BAL)
35   1  1  #       (COL(1),A(20),COL(25),F(10),COL(40),F(10,2));
36   1  1  #     ENDDO MASTER_LOOP
37   1  1  #     PUT EDIT ('***END OF DATA ***')               /* TRAILER*/
38   1  1  #     (SKIP(6), A);
           #                                                   /*1  HEADING PROC */
39   2  2  #     HEADING: PROCEDURE;
40   2  2  #       PUT EDIT('CHARGE ACCOUNT TRANSACTION SUMMARY',
41   2  2  #       'NAME','ACCOUNT','NEW BAL.','CHARGE','PAYMENTS')
42   2  2  #       (COL(20),A,COL(4),A,COL(28),A,COL(44),A,COL(58),A,COL(68),A)
43   2  2  #       ;
44   2  1  #     ENDPROC HEADING
45      1  #   ENDPROC CHARGE
```

(b)

process of pseudo-coding to translate programming requirements into structured code constructs is reasonably clear.

Notes and References

1. Hughes, Joan and Michton, Jay. *A Structured Approach to Programming.* Englewood Cliffs, N.J.: Prentice-Hall, 1977.
2. Van Leer, P. "Top Down Development Using a Program Design Language." IBM Systems Journal, vol. XV, no. 2, 1976, pp. 155–170.
3. McGowan and Kelly. *Top-Down Structured Programming Techniques.* New York: Petrocelli Books, 1975.
4. Yourdan, Edward. *Techniques of Program Structure and Design.* Englewood Cliffs, N.J.: Prentice-Hall, Inc., 1975.

SUPPLEMENT TO CHAPTER 6
ABC Auto Parts—A Case Study

Structured program design is a framework for solving system generation problems. The basic tools—hierarchical tree structures, HIPO charts, and pseudo-code—find usefulness in all aspects of the design, implementation, and maintenance phases of a productive system. In the preceding chapters, we have approached these subjects individually. Now we will show how they are integrated into a design philosophy. For this exercise, we will introduce ABC Auto Parts for problem analysis. While the case does not contain the complexity and intricacy of large systems, it serves as a model illustrating important concepts essential in a structured design philosophy.

INPUT/OUTPUT REQUIREMENTS

ABC Auto Parts consists of three retailing stores. Weekly activity reports are produced with a weekly summary of inventory movement. The report includes the following information:

1. Total sales and returns for each part sold
2. Intermediate totals for each store and all the parts sold
3. Final sales totals for all stores

A program is needed to provide the weekly summary of inventory

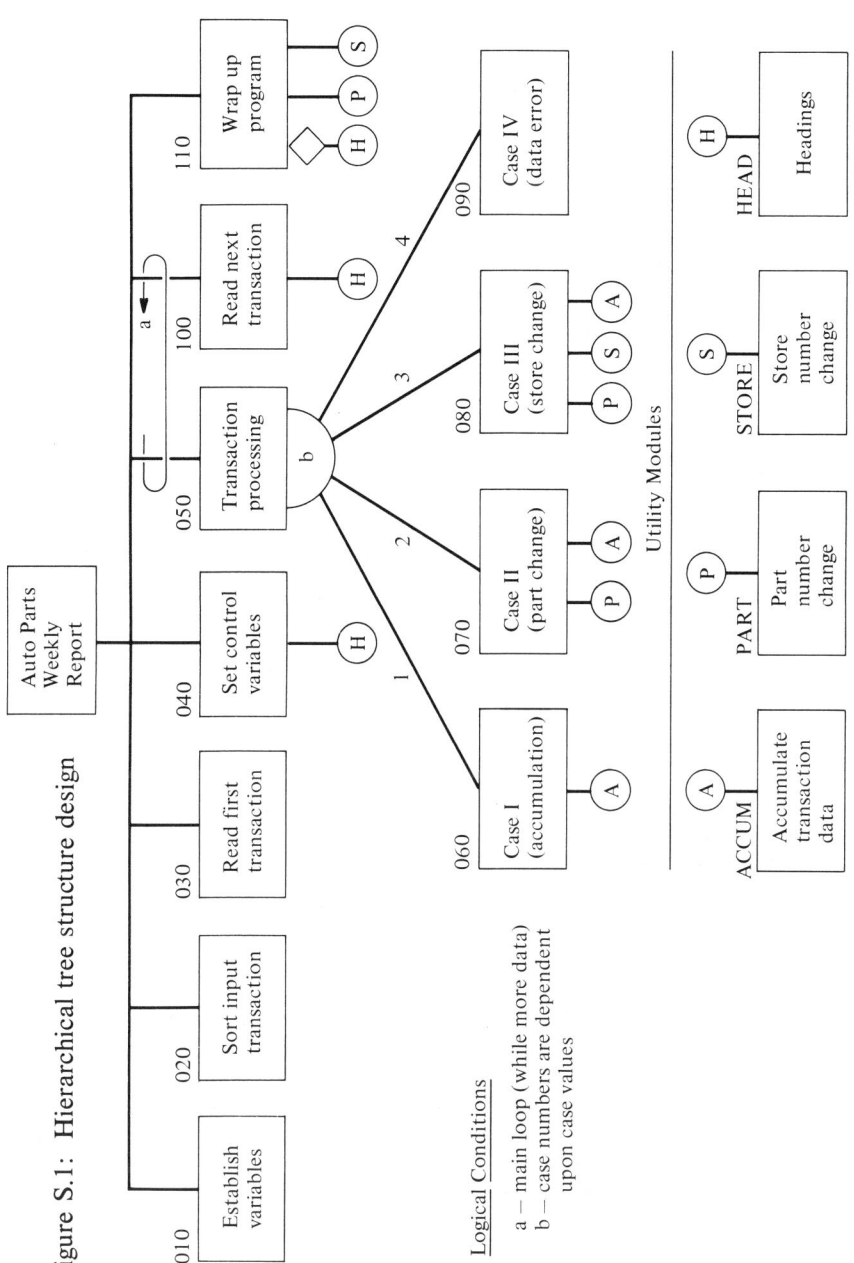

A Primer on Structured Program Design

movement. The data for each inventory transaction will include the following information:

1. Store number
2. Part number
3. Number of units sold
4. Price per unit
5. Control character
   a. '1' for sale
   b. '2' for return

Develop the following for this problem:

1. Tree structure design
2. Pseudo-code
3. Resulting PL/I program

Tree Structure Design

After some consideration of the logical requirements for this problem, the tree structure shown in Figure S.1 can be constructed. A brief functional description of these logical segments is given in Table S.1.

Pseudo-Code

Figure S.2 outlines the macro-level pseudo-code corresponding to the tree structure design. Key emphasis at this point is on loop control and placement of major logical segments. Notice that the various tree structure segments are identified in the source code using right-hand margin comments.

Logic Development

There are three key logical segments necessary to properly handle the data:

1. Accumulation of common part number data (sales and returns)
2. Handling of part number changes
3. Handling of store number changes

In order to execute properly, store number changes must take precedence over part number changes.

Accumulation. Three subunits of logic are implicit in handling accumulations of units sold. A control character value of 1 implies a sale, while a 2 indicates return of the item. Any other value is invalid. The

Pseudo-Coding

Table S.1. Functional Segments Description (Related to Figure S.1)

Segment	Description
010	INITIALIZE AND DECLARE: Describes input and output, sets up files, and establishes end-of-file condition
020	SORT INPUT TRANSACTION: Sorts input file into this program
030	READ FIRST TRANSACTION: Opens files and reads the first transaction record
040	SET CONTROL VARIABLES: Sets the controls for store number and part number using the first card read and prints the heading
050	TRANSACTION PROCESSING: Processes transaction according to the computed case value
060	CASE I: Accumulates part information
070	CASE II: Prints part total line (OUTPUT-LINE) and accumulates a new part
080	CASE III: Prints part total line (OUTPUT-LINE), prints store total line (TOT-LINE) and accumulates a new part
090	CASE IV: In case of data error this segment prints an error message (CASE-ERROR)
100	READ A TRANSACTION: Reads another transaction record and prints heading if line max is exceeded
110	WRAP UP PROGRAM: Prints headings if line max is exceeded, prints part total line (OUTPUT-LINE), prints store total line (TOT-LINE), prints grand total line (GRAND-TOT-LINE) and closes file

expanded pseudo-code to accomplish this is shown in Figure S.3.

Part Number Change. Whenever a part number change occurs all sales and return totals must be written, then store sales totals accumulated. Part number accumulators are then zeroed out in preparation for subsequent items. Following this the same accumulation logic as described above is required. The pseudo-code for handling this is shown in Figure S.4.

Figure S.2. Pseudo-Code for ABC Case Study (Version I)

/* ABC AUTO PARTS WEEKLY REPORT */
ABC PROC MAIN
   DATA
/* INITIALIZE AND DECLARE */   /* SEGMENT 010 */
   'describe input records'
   'describe output lines'
   'set up files'
   'set up variables'
   'establish endfile condition'
/* SORT INPUT TRANSACTION */ 'sort input file'/* SEGMENT 020 */
   'sort input file'
/* READ FIRST TRANSACTION */   /* SEGMENT 030 */
   'open files'
   'read first card'
/* SET CONTROL VARIABLES */   /* SEGMENT 040 */
   'set store control'
   'set part control'
   'print heading'
/* ITERATIVE LOOP FOR TRANSACTION */ /* SEGMENT 050 */
   MAINLOOP:
      DO WHILE ('more data')
         'initialize case value'
         CASEENTRY
      CASE(1): 'accumulate transaction'   /* SEGMENT 060 */
      CASE(2): 'part number change'   /* SEGMENT 070 */
      CASE(3): 'store number change'   /* SEGMENT 080 */
      CASE(4): 'transaction error'   /* SEGMENT 090 */
         ENDCASE
/* READ A TRANSACTION */   /* SEGMENT 100 */
      'read a card'
   END MAINLOOP
/* WRAP UP PROGRAM */   /* SEGMENT 110 */
   'program wrapup logic'
END ABC

Store Number Change. Three major subunits of logic are necessary to accomplish a store number change: the part number logic must be executed; then store totals created; finally, the accumulation logic is needed. Pseudo-code for the store portion of this logic segment is shown in Figure S.5.

Figure S.3:   Accumulation Pseudo-Code

```
ACCUM: PROCEDURE;
   IF 'transaction code error' THEN
      'print error message'
   ENDIF
   IF 'control character = 1' THEN
      'accumulate sales totals'
   ELSE IF 'control character = 2' THEN
      'accumulate returns totals'
         ENDIF
   ENDIF
END ACCUM;
```

Figure S.4:   Part Number Change of Pseudo-Code

```
PART: PROCEDURE
   'print part number accumulated totals'
   'increment printer line number'
   'accumulate store sales in units'
   'accumulate store sales in dollars'
   'accumulate store returns in units'
   'accumulate store returns in dollars'
   'zero accumulators'
   'set part control'
END PART
```

Figure S.5:   Store Number Change Pseudo-Code

```
STORE: PROCEDURE
   'print store summaries'
   'accumulate store sales in units'
   'accumulate store sales in dollars'
   'accumulate return sales in units'
   'accumulate return sales in dollars'
   'set store control'
END STORE
```

Summary of Transaction Logic

In CASE(1) the accumulation logic is executed by the statement

>CALL ACCUM

Part number changes, CASE(2), require both the PART and ACCUM logic be executed in the sequence

>CALL PART
>CALL ACCUM

Finally, store number changes are handled by the following combination:

>CALL PART
>CALL STORE
>CALL ACCUM

One last subtle design point involves the end of file condition. A review of Figure S.1 shows that the last record should be read within segment 100 which means that the wrapup segment 110 must perform final part and store accumulations before printing end of report summaries. Failure to do this would omit the last record from summary results.

This example was created because it is similar to many report generation programs found in commercial applications. Note here that two major break levels are implicit in the data. These are analogous to subtotals (parts) and final totals (stores) for a report.

Figure S.6 shows the source program resulting from the design outlined in figures S.1 through S.5. This program is a natural extension of the pseudo-code discussed thus far. The reader should spend some time dissecting the final program. Notice that the general logic segments for accumulation, part changes, store changes, and heading generation were found to be used in multiple program segments. For these reasons they are each organized as subroutines and called when needed. Use of the approach increases execution time due to subroutine invocation overhead, however improvement in program structure and readability is felt to outweigh the disadvantages. Figure S.7 contains the input for the example program, and Figure S.8 is a listing of the generated output.

/*H ABC AUTO PARTS -- CASE STUDY */

Figure S.6: ABC Auto Parts program source code

```
STMT  LEV  NEST           SOURCE PROGRAM TEXT

               #   /*H  ABC AUTO PARTS -- CASE STUDY  */
  1    1         #   ABC: PROC OPTIONS(MAIN);
  2    1         #     DECLARE
                 #
                 #     /***** 010 INITIALIZE AND DECLARE *****/
                 #
                 #     /*S010****/
                 #
                 #     /* ******** INPUT RECORD ******** */
                 #       1 CARD,
                 #         2 STORE_NUMBER              PICTURE  '9999',
                 #         2 PART_NUMBER               PICTURE  '999999',
                 #         2 NUMBER_OF_UNITS           PICTURE  '999',
                 #         2 PRICE_PER_UNIT            PICTURE  '999V99',
                 #         2 FILLERA                   CHAR(61),
                 #         2 CONTROL_CHARACTER         PICTURE  '9',
                 #
                 #     /* ******** PART OUTPUT LINE ******** */
                 #       1 PART_LINE,
                 #         2 CC                        CHAR(1)   INIT (' '),
                 #         2 FILLA                     CHAR(5)   INIT (' '),
                 #         2 CONTROL_STORE_NUMBER      PICTURE  'ZZ9'    INIT (0),
                 #         2 FILLB                     CHAR(5)   INIT (' '),
                 #         2 CONTROL_PART_NUMBER       PICTURE  '999999' INIT (0),
                 #         2 FILLC                     CHAR(7)   INIT (' '),
                 #         2 PART_SALES_UNITS          PICTURE  'ZZ9'    INIT (0),
```

/*H ABC AUTO PARTS -- CASE STUDY */

STMT LEV NEST   SOURCE PROGRAM TEXT

```
#                    2 FILLD                   CHAR(3)   INIT (' ').
#                    2 PART_SALES_AMOUNT       PICTURE '$$$$9V.99'  INIT (0).
#                    2 FILLE                   CHAR(6)   INIT (' ').
#                    2 PART_RETURN_UNITS       PICTURE 'ZZ9'  INIT (0).
#                    2 FILLF                   CHAR(4)   INIT (' ').
#                    2 PART_RETURN_AMOUNT      PICTURE '$$$$9V.99'  INIT (0).
#                    2 REST                    CHAR(71)  INIT (' ').
#
/* ********* STORE TOTAL OUTPUT LINE ********* */
#             1 TOT_LINE,
#                    2 CC                      CHAR(1)   INIT (' ').
#                    2 LITERAL                 CHAR(25)  INIT ('TOTALS').
#                    2 STORE_SALES_UNITS       PICTURE 'ZZZ9'  INIT (0).
#                    2 FILLA                   CHAR(2)   INIT (' ').
#                    2 STORE_SALES_AMOUNT      PICTURE '$$$$9V.99'  INIT (0).
#                    2 FILLB                   CHAR(5)   INIT (' ').
#                    2 STORE_RETURN_UNITS      PICTURE 'ZZZ9'  INIT (0).
#                    2 FILLC                   CHAR(3)   INIT (' ').
#                    2 STORE_RETURN_AMOUNT     PICTURE '$$$$9V.99'  INIT (0).
#                    2 REST                    CHAR(71)  INIT (' ').
#             1 BLANK_LINE,
#                    2 CC                      CHAR(1)   INIT ('0').
#                    2 REMAINDER               CHAR(132) INIT (' ').
/* ********* GRAND TOTAL OUTPUT LINE ********* */
#             1 GRAND_TOT_LINE,
#                    2 CC                      CHAR(1)   INIT ('0').
#                    2 LITERAL                 CHAR(24)  INIT (
                       'GRAND TOTALS                         ').
```

```
   #          2 FINAL_SALES_UNITS                    PICTURE 'ZZZZ9'    INIT (0).
   #          2 FILLA                                CHAR(3)  INIT (' ').
   #          2 FINAL_SALES_AMOUNT                   PICTURE '$$$$9V.99'  INIT (0).
   #          2 FILLB                                CHAR(4)  INIT(' ').
   #          2 FINAL_RETURN_UNITS                   PICTURE 'ZZZZ9'    INIT (0).
   #          2 FILLC                                CHAR(4)  INIT (' ').
   #          2 FINAL_RETURN_AMOUNT                  PICTURE '$$$$9V.99'  INIT (0).
   #          2 REST                                 CHAR(71) INIT (' ').
   #   /* ********* CASE ERROR OUTPUT LINE ********* */
   #          1 CASE_ERROR,
   #            2 CC                                 CHAR(1)  INIT ('0').
   #            2 MSG                                CHAR(43) INIT
   #   (' ERROR IN CASE VALUE-CHECK DATA.CASE VALUE=').
   #            2 CV                                 PICTURE '9999'.
   #            2 REST                               CHAR(85).
   #   /* ********* FILE DECLARATION ********* */
   #   CARDIN  FILE INPUT  RECORD ENV(F(80)).
   #   PRINTER FILE OUTPUT RECORD ENV(F(133)) CTLASA).
   #   /* ********* VARIABLE DECLARATION ********* */
   #   CASE(4)                                       LABEL.
   #   DOLLAR_AMOUNT                                 PICTURE '$$$$99.99'.
   #   DONE                                          BIT(1)   INIT ('0'B).
   #   LINE_NO                                       FIXED    INIT (55).
   #   LINE_MAX                                      FIXED(3) INIT (55).
   #   PRINT_LINE                                    CHAR(133) INIT (' ');
 3 1 ON ENDFILE(CARDIN) DONE = '1'B;
   # /*S020****/
   #
```

/*H ABC AUTO PARTS --- CASE STUDY */

```
STMT  LEV NEST              SOURCE PROGRAM TEXT

                   *      /*****  020  SORT_INPUT_TRANSACTION  *****/
                   *      /* ***  NOT INCLUDED IN THIS PROGRAM * ***/
                   *
                   *      /*S030*****/
                   *
                   *      /*****  030 READ FIRST TRANSACTION RECORD *****/
   4    1          *      OPEN FILE (CARDIN), FILE (PRINTER);
   5    1          *      READ FILE (CARDIN) INTO (CARD);
                   *
                   *      /*S040*****/
                   *
                   *      /*****  040 SET CONTROL VARIABLES  *****/
   6    1          *      CONTROL_STORE_NUMBER = STORE_NUMBER;
   7    1          *      CONTROL_PART_NUMBER = PART_NUMBER;
   8    1          *      IF LINE_NO >= LINE_MAX THEN CALL HEAD;
                   *
                   *      /*S050*****/
                   *
                   *      /*****  050 ITERATIVE LOOP FOR TRANSACTION *****/
   9    1          *      MAIN_LOOP:
  11    1          *      DO WHILE (¬DONE);
                   *      /* *** FOR EACH INFILE RECORD * ***/
  12    1   1      *         IF CONTROL_STORE_NUMBER¬=STORE_NUMBER
  13    1   1      *            THEN CASE_VALUE=3;
  14    1   1      *         ELSE
  15    1   1      *            IF CONTROL_PART_NUMBER> PART_NUMBER
  16    1   1      *               THEN CASE_VALUE=4;
  17    1   1      *            ELSE
```

134

```
18   1  1  *            IF CONTROL_PART_NUMBER=PART_NUMBER
19   1  1  *                THEN CASE_VALUE=1;
20   1  1  *            ELSE
21   1  1  *                CASE_VALUE=2;
22   1  1  *            GO TO CASE(CASE_VALUE);
           *        /* *** TRANSFER TO SELECTED CASE LOGIC * ***/
           *
           *        /*S060*****/
           *
           *        /****** 060 CASE I  *****/
           *        /* CASE 1 SEGMENT: PART NUMBER ACCUMULATION */
23   1  1  *        CASE(1):
24   1  1  *            CALL ACCUM;
25   1  1  *            GO TO END_CASE;
           *
           *        /*S070*****/
           *
           *        /****** 070 CASE II  *****/
           *        /* CASE 2 SEGMENT: PART NUMBER LINE TOTALS */
26   1  1  *        CASE(2):
27   1  2  *        DO;
28   1  2  *            CALL PART;
29   1  2  *            CALL ACCUM;
30   1  2  *            GO TO END_CASE;
31   1  1  *        END;
           *
           *        /*S080*****/
           *
           *        /****** 080 CASE III *****/
           *        /* CASE 3 SEGMENT: STORE LINE TOTALS */
32   1  1  *        CASE(3):
33   1  2  *        DO;
```

/*H ABC AUTO PARTS -- CASE STUDY */

STMT	LEV	NEST		SOURCE PROGRAM TEXT
34	1	2	#	CALL PART;
35	1	2	#	CALL STORE;
36	1	2	#	CALL ACCUM;
37	1	2	#	GO TO END_CASE;
38	1	1	#	END;
			#	/*S090*****/
			#	
			#	/***** 090 CASE IV *****/
			#	/* CASE 4 SEGMENT: ERROR CHECK FOR CASE_VALUE */
39	1	1	#	CASE(4):
40	1	1	#	WRITE FILE (PRINTER) FROM (CASE_ERROR);
			#	
			#	/* END_CASE */
41	1	1	#	END_CASE:
42	1	1	#	;
			#	
			#	/*S100*****/
			#	
			#	/***** 100 READ A TRANSACTION CARD *****/
43	1	1	#	READ FILE (CARDIN) INTO (CARD);
44	1	1	#	IF LINE_NO >= LINE_MAX THEN CALL HEAD;
45	1	1	#	END MAIN_LOOP;
			#	/*S110*****/
			#	
			#	/***** 110 WRAPUP PROGRAM *****/
46	1	1	#	IF LINE_NO >= LINE_MAX THEN CALL HEAD;
47	1	1	#	CALL PART;

```
48  1  *   CALL STORE;
49  1  *   WRITE FILE (PRINTER) FROM (GRAND_TOT_LINE);
50  1  *   WRITE FILE (PRINTER) FROM (PRINT_LINE);
51  1  *   CLOSE FILE (CARDIN), FILE (PRINTER);
        *
        *   /*  U T I L I T Y   M O D U L E S  */
        *
        *   /************    A    ************/
52  2  *   ACCUM: PROCEDURE;
        *   /* *** PERFORM PART ACCUMULATION *** */
53  2  1   IF CONTROL_CHARACTER ¬= 1 & CONTROL_CHARACTER ¬= 2 THEN DO;
55  2  1       PRINT_LINE = '0'||(30)' '||'CONTROL CHARACTER ERROR' ;
56  2  1       WRITE FILE(PRINTER) FROM (PRINT_LINE);
57  2  1       LINE_NO = LINE_NO + 1;
58  2  1       PRINT_LINE = ' ';
59  2  1                                                           END;
60  2  1   IF CONTROL_CHARACTER = 1 THEN DO;
62  2  1       DOLLAR_AMOUNT = NUMBER_OF_UNITS * PRICE_PER_UNIT;
63  2  1       PART_SALES_UNITS = PART_SALES_UNITS + NUMBER_OF_UNITS;
64  2  1       PART_SALES_AMOUNT = PART_SALES_AMOUNT + DOLLAR_AMOUNT;
65  2  1                                                           END;
66  2      ELSE IF CONTROL_CHARACTER = 2 THEN DO;
69  2  1       DOLLAR_AMOUNT = NUMBER_OF_UNITS * PRICE_PER_UNIT;
70  2  1       PART_RETURN_UNITS = PART_RETURN_UNITS + NUMBER_OF_UNITS;
71  2  1       PART_RETURN_AMOUNT = PART_RETURN_AMOUNT + DOLLAR_AMOUNT;
72  2  1                                                           END;
73  2      END ACCUM;
        *   /************    P    ************/
74  2  *   PART: PROCEDURE;
        *   /* *** OUTPUT PART SUMMARY AND PERFORM STORE ACCUMULATION *** */
75  2  *   WRITE FILE (PRINTER) FROM (PART_LINE);
```

/*H ABC AUTO PARTS -- CASE STUDY */

STMT	LEV NEST		SOURCE PROGRAM TEXT
76	2	#	LINE_NO = LINE_NO + 1;
77	2	#	STORE_SALES_UNITS = STORE_SALES_UNITS + PART_SALES_UNITS;
78	2	#	STORE_SALES_AMOUNT = STORE_SALES_AMOUNT + PART_SALES_AMOUNT;
79	2	#	STORE_RETURN_UNITS = STORE_RETURN_UNITS + PART_RETURN_UNITS;
80	2	#	STORE_RETURN_AMOUNT = STORE_RETURN_AMOUNT + PART_RETURN_AMOUNT;
81	2	#	PART_SALES_UNITS,PART_SALES_AMOUNT,PART_RETURN_UNITS,
82	2	#	PART_RETURN_AMOUNT = 0;
83	2	#	CONTROL_PART_NUMBER = PART_NUMBER;
84	1	#	END PART;
		#	
		#	/********* S **********/
85	2	#	STORE: PROCEDURE;
		#	/* *** OUTPUT STORE SUMMARY AND PERFORM FINAL ACCUMULATION *** */
86	2	#	WRITE FILE (PRINTER) FROM (TOT_LINE);
87	2	#	WRITE FILE (PRINTER) FROM (BLANK_LINE);
88	2	#	LINE_NO = LINE_NO + 4;
89	2	#	FINAL_SALES_UNITS = FINAL_SALES_UNITS + STORE_SALES_UNITS;
90	2	#	FINAL_SALES_AMOUNT = FINAL_SALES_AMOUNT + STORE_SALES_AMOUNT;
91	2	#	FINAL_RETURN_UNITS = FINAL_RETURN_UNITS + STORE_RETURN_UNITS;
92	2	#	FINAL_RETURN_AMOUNT = FINAL_RETURN_AMOUNT + STORE_RETURN_AMOUNT;
93	2	#	STORE_SALES_UNITS,STORE_SALES_AMOUNT,STORE_RETURN_UNITS,
94	2	#	STORE_RETURN_AMOUNT = 0;
95	2	#	CONTROL_STORE_NUMBER = STORE_NUMBER;
96	1	#	END STORE;
		#	
		#	/********* H **********/
97	2	#	HEAD: PROCEDURE;

```
#       /* *** CREATE HEADINGS * ***/
#       PRINT_LINE =.1.||(19).  .||.ABC AUTO PARTS INC..  ;
#       WRITE FILE (PRINTER) FROM (PRINT_LINE);
#       PRINT_LINE=.0.||(16).  .||.WEEKLY INVENTORY MOVEMENT.  ;
#       WRITE FILE (PRINTER) FROM (PRINT_LINE);
#       PRINT_LINE =.0.||(132).  .  ;
#       WRITE FILE (PRINTER) FROM (PRINT_LINE);
#       PRINT_LINE = .0.||(29).  .||.SALES.||(15).  .||.RETURN.;
#       WRITE FILE (PRINTER) FROM (PRINT_LINE);
#       PRINT_LINE=(15).  .||.PART.||.                          .||
#         .                          ----- ;
#       WRITE FILE (PRINTER) FROM (PRINT_LINE);
#       PRINT_LINE = .0.||.   STORE.||.       NUMBER.||.     UNITS.||
#         .  AMOUNT.||.   UNITS.||.    AMOUNT.;
#       WRITE FILE (PRINTER) FROM (PRINT_LINE);
#       PRINT_LINE =.                          -----  -----
#         .||.                          -----  -----     ;
#       WRITE FILE (PRINTER) FROM (PRINT_LINE);
#       PRINT_LINE = .  .;
#       LINE_NO = 9;
#   END HEAD;
#   END ABC;
```

Figure S.7: ABC Auto Parts input data listing

```
( ABC CASE STUDY DATA )                 1
085100003203001000                      1
085100004201501000                      1
085100004601700706                      1
085100004601701483                      1
085100004601702296                      2
085100004601701395                      2
085100004601701500                      1
085100004701201195                      2
085100004701201295                      1
085100005700500999                      1
085100005700501000                      1
085100005700501332                      1
085100005700501625                      1
085100005700501986                      1
085100005700502342                      1
085100005700502451                      1
085100005700503150                      1
085100005700503222                      1
085100005700503841                      1
085100005700504083                      1
085100005700504533                      1
085100005700505968                      1
085100005700506998                      2
085100005700500895                      2
085100005700501125                      2
085100005700501382                      2
085100005700501445                      2
085100005700501529                      2
085100005700501896                      2
085100005700502103                      2
085100005700502448                      2
085100005700503030                      2
085100005700503742                      1
085100006503001995                      2
085100006503001495                      1
085200001301705207                      1
085300001301707593                      1
085300002401509310                      1
085300004602501003                      1
085300004602501475                      1
085300004602501528
```

Pseudo-Coding

Figure S.8: ABC Auto Parts program output

ABC AUTO PARTS INC.

WEEKLY INVENTORY MOVEMENT

		SALES		RETURN	
STORE	PART NUMBER	UNITS	AMOUNT	UNITS	AMOUNT
851	000032	30	$300.00	0	$0.00
851	000042	15	$150.00	0	$0.00
851	000046	51	$762.00	34	$492.00
851	000047	12	$143.00	12	$155.00
851	000057	70	$2171.00	50	$976.00
851	000065	30	$598.00	30	$448.00
TOTALS		208	$4124.00	126	$2071.00
852	000013	17	$885.00	0	$0.00
TOTALS		17	$885.00	0	$0.00
853	000013	17	$1290.00	0	$0.00
853	000024	15	$1396.00	0	$0.00
853	000046	75	$1000.00	0	$0.00
TOTALS		107	$3686.00	0	$0.00
GRAND TOTALS		332	$8695.00	126	$2071.00

7

PROGRAM STYLE AND DEBUGGING

Since the advent of computer applications, the process of developing a program has been considered to be primarily an art. Design techniques, readability, and code structures were often free form and unique for each application. However, the structured programming revolution is focusing increased attention on programming style and standards. Two key conclusions have emerged:

1. Software development and maintenance costs are becoming increasingly costly and techniques are needed to manage these aspects of programming.
2. Readability is more important than efficiency (10% of codes typically consume 50% or more of execution time).

For these and many other related reasons the approach to improving program style is of increased importance to the contemporary programmer. This chapter describes a series of suggested procedures and practices which, if followed and used in good faith, should lead to a reduction in program development time and related errors.

Our goal here is to outline some common programming design problems. In this discussion we are addressing the very issue of the process itself, e.g., What is its goal? One response to this question might be to write *good* programs (who could argue with such a goal as this?). Since there are so many perceptions as to what this entails, let us summarize eight desirable attributes of a good program and its related design process:

1. The final product performs a useful function accurately.
2. Whenever the program fails, it does so in a safe manner causing minimal operational disruption.
3. Testing costs are minimized by a coherent design structure.

4. Subsequent maintenance costs are decreased by rigorous code structures and style conventions.
5. Future modification opportunity is facilitated by the design structure.
6. Development cost is planned to meet organizational constraints and correspondingly controlled throughout the design cycle.
7. The design is kept simple and easy to comprehend.
8. Special efficiency considerations are implemented only to the required degree, realizing that many other attributes can be sacrificed by pure efficiency changes.

In summary, the philosophy of good program design as used here suggests that it be simple, functional, and not filled with efficiency tricks which cloud the logic. Program style considerations aid in the orderly development of good programs. The remainder of this chapter will discuss in some detail key issues of style and debugging.

PROGRAMMING STYLE

Programming style refers to the conscious acceptance and internalization on the part of the programmer of certain standards. "If there is more than one way to do something and the choice is rather arbitrary, pick one way and always do it that way."[14] If standards are adhered to, the result will be more precise and understandable communication in the programs written. Only when the standard is too restrictive or too cumbersome should it be ignored. Much of the problem of standardization involves programmers' egos and their unique approaches to the "best" way. Beware of programmers who ignore set standards because they do not want to change their own style or who do not wish to rewrite bad segments of old programs. The basic elements of programming style include expression, structure, documentation and readability, guarding against common blunders, input/output considerations, and efficiency.

Expression

Expression is the mode, or means, by which a programmer uses words, phrases, or symbols to significantly represent a desired idea, thought, or action. In other words, in programming it is necessary to say exactly what you mean in the program.[8] There are many standards that should be adhered to for quality in expression.

Good expression requires avoidance of temporary variables. Many times an inexperienced programmer will include excessive intermediate variables in a program. Such variables tend to cloud the algorithms being used and their use should be carefully examined in each case. There is generally an issue of efficiency to be considered when using temporary variables. Items which reappear frequently obviously justify saving partial results for later use. The guideline here is that clarity should not be sacrificed for efficiency.

Experienced programmers often tend to get overly involved in a program at the bit level. Those who try to work on the bit, half-byte, or byte level often destroy the readability of a program. This type of logic development should be avoided unless core availability is restricted. The contemporary philosophy is that the compiler should be allowed to handle the dirty work of variable size and allocation wherever possible.[8]

If a certain function is used repeatedly in a program, the readability of the program can be enhanced by using a call invocation to a common procedure and coding the function in the program only once.

All programmers should use standard variable naming conventions. By using standard abbreviations it is less likely that several different abbreviations for the same variable will exist. Use of standardized abbreviations by all programmers in the organization also makes maintenance a much easier task.[14] The U.S. Air Force, for example, publishes a list of commonly used internal variables under the title *Standard Data Elements*.

Programmers, in an attempt to "get the program out fast," often hurry through their program logic development. Numerous conditional branches are often included in a program which would have been unnecessary if the programmer had just spent a little more time in the logical formulation of the problem. Extra conditional branches should never be substituted for sound logical expression. A logical expression which is vague or hard to understand should never be left in a program. It can always be reworded into a more understandable form.[8]

Internal Documentation and Readability

The element of software documentation is closely related to expression, since it helps to make a program more readable and understandable. Documentation can also be of great help in error identification. The most common form of internal documentation is comment statements embedded in the source. Use of indentation and blank source lines to show logical segments of code in the program make logic structure easier to see and understand, just as paragraphs make compositions easier to read and understand.[8] Comment cards which are used before the beginning of a program segment to describe the segment and give pertinent facts and details about it are called prologue comments. Explanatory comments are nested inside the program segments. They explain any code that is too complex to be explained in the prologue or any code that is not self-explanatory.

Not just any comment is good, meaning certain standards for comments must be met. Comments and code must always agree; comments should not be a restatement of the code but, instead, a definitive statement of the code. All comments should be important, or they will be discounted by the reader. Too many comments also detract from readability. While comments are important, over-commenting is harmful.[8]

A second form of subtle internal documentation is the use of blank lines. Just as blank lines between logic segments and the use of indentation to highlight segments of code were good for expression, so also are they good for readability and documentation. They help delineate program structure.

Four different types of comments can be used for internal documentation. In decreasing order of importance these are:

1. Procedure or segment description
2. Logical module delineation
3. Notes on beginning and end of key logic segments
4. Miscellaneous notes

Items 1 and 2 are quite useful in improving readability. Conversely, items 3 and 4 should not be done unless the code structure is nonstandard or unusually complex. Good code is self-documenting. It should be realized that there is considerable disagreement over an appropriate level of commenting. Most programmers err in the direction of too few, rather than too many.

A second form of internal documentation is indentation of the source code to highlight logical structures. Several programs are available to mechanically format source listings to improve readability. Chapter 6 illustrated this aspect of program documentation, so we shall not dwell on it here.

A third method to improve internal documentation is the use of "white space" to highlight and group selected portions of the source

Program Style and Debugging

listing. Page skipping aggregates logical segments onto separate pages. In this manner, using the "GOTO-less" figures of structured programming, errors can be isolated to one- or two-page listings rather than spread over an entire program.

The fourth and final form of documentation is the proper selection of variable names used in the program. In a production environment variable names should be standardized and published in a common data element dictionary. Certain other locally used variables may not need to follow such rigorous naming conventions, however the names used should be descriptive of their function. Module (procedure) and loop labels should follow some standard naming convention which describes their function.

Structure

Use of good programming conventions makes a program more logical, more readable, and easier to rework. Many of the elements of structured programming are related to this aspect of program design. For example, logically organizing variables into aggregates such as arrays (tables) or structures (records in COBOL) may simplify the structure of a program. By using arrays, complex repetitive control sequences can often be avoided. In general, when the programmer has control over representation of data, he/she should choose one that will make the program simpler. Data representation will be discussed further in chapter 8.

The programmer should learn to design programs using a pseudo-language and then translate it into code in the manner described in chapter 6. When the basic logic units of the pseudo-language are applied to a problem, the designer tends to think in purely logical terms and not in terms of a programming language. The result is a more clearly developed and logically structured program.

In order to keep the structure logical, a programmer must be prepared to rewrite bad code. Bad code that is patched distorts the structure and readability of a program and will usually return to haunt the process during debugging or future maintenance. Related to the idea of modularity in programming is the concept of writing and testing a big program in small pieces. By working with small, logical segments of a program, it is possible to get a better grasp on the problem and solve it gradually, worrying only with establishment of proper interfaces. Some call this process "eating a dinosaur" (e.g., tough and hard to swallow, but possible in small pieces).

Common Blunders and Their Prevention

There are certain types of mistakes with which almost every programmer has trouble.[7] This section presents an overview of some of these and provides suggestions as to what to look for and how to prevent them.

The most common error found in a program is that of uninitialized variables and constants.[8] The uninitialized constant error is the easier to fix, since all that is required is to develop the practice of initializing all constants at the beginning of the program. The initialization of variables can be more troublesome, but is generally looked at as housekeeping which should be done early in the program. Often this is somewhat more difficult to recognize as an error, since in some cases variables need to be initialized in more than one place in the program. Care must be taken to insure that values are assigned to variables and constants within the proper logical segment, then reset where appropriate.

The second common mistake is the "off-by-one" error. This occurs most often where transfers are to occur on a certain condition. In order to be sure that the condition specified is the desired one, the programmer should test the segment with a small set of mock data through more than one cycle of the program. A related error is the execution of a wrong branch for a conditional segment at boundary conditions. For example, the instruction IF $X=0$ might be used erroneously instead of IF $X>0$. In this case a test for the zero condition would branch to the wrong set of logic.

Input/Output Considerations

Proper design of input and output will make programs less susceptible to bugs. The programmer can make input easier to prepare by using a uniform input format. This also makes the input easier to proofread. When possible, free-form input should be used. Free-form input and the use of separation fields (or delimiters) can have many headaches caused by keypunching errors. The use of self-identifying input may also be an advantage when a small amount of data is required.

Once the format of the input has been established, a data sample should be tested to make sure that it does not exceed the limits set in the program. The input should be logically terminated by an end-of-file marker and not by a preset counter. Bad data discovered by the program should always be flagged and a restart attempted wherever possible.

Some general form of data editing is desirable to help "purify" a program. Tactics such as testing for excessive values, illegal codes, zero values, or other logical tests can save catastrophic problems later. During program testing all input values should be rewritten to an output device in order to allow user verification of the input stream. This isolates problems to the program itself, rather than allowing undefined external problems to be involved. Output of the program should be well formatted and self-explanatory, plus all bad data captured during execution should be printed out for inspection.

Other Style Considerations

Several other style elements should also be mentioned here. Even though they are listed briefly, their importance to the overall program style should not be reduced:

1. All logical blocks should be labeled, beginning and end.
2. Logical blocks should be individually closed.
3. Each procedure will either contain its own error-handling routine or reference via comment to identify which external procedure is used.
4. Key word abbreviations should be either avoided completely or used exclusively. Don't mix spellings.
5. Each procedure should be assigned a name reflecting its function and a brief explanation in comment form.
6. When GOTO's are used outside of error-handling logic, they must follow one of two rules:
   a. Branch to the first statement following their block.
   b. Never branch into a block segment or never branch backwards.

Efficiency

Efficiency is covered last as an element of programming style because it is the subordinate element. The program must, first, perform its function and, second, be clear and easy to follow. Only then should efficiency be addressed. The traditional programmer was concerned with efficiency as a direct function of the particular machine on which the program ran. Contemporary programmers must now be concerned with building programs that can run on many machines and outlast a single generation of machines. After this is accomplished, they should be concerned with machine-dependent optimization.[14] Clarity and

Table 7.1. The Laws of Programming

Definition:	A "working" program is one that has only unobserved bugs.
Law I:	Every nontrivial program has at least one bug. Corollary I: A sufficient condition for program triviality is that it has no bugs. Corollary II: At least one bug will be observed after the author leaves the organization.
Law II:	The subtlest bugs cause the greatest damage or problems. Corollary I: A subtle bug will modify storage, thereby masquerading as some other problem.
Law III:	Bugs will appear in one part of a working program when an "unrelated" part is modified.
Law IV:	*Lulled into Security Law.* A "debugged" program that crashes will wipe out source files on storage devices when there is the least available backup.
Law V:	A hardware failure will cause system software to crash, and the CE (computer company's custom engineer) will blame the programmers.
Law VI:	A system software crash will cause hardware to act strangely and the programmers will blame the CE.
Law VII:	The documented interfaces between standard software will have undocumented quirks.
Law VIII:	The probability of a hardware failure disappearing is inversely proportional to the distance between the computer and the CE.
Law IX:	Murphy designed the computer in your organization.
Law X:	O'Shea's Law: Murphy was an optimist. (Note: For those who may be in doubt, Murphy's law states that if anything can go wrong, it will, and always at the most inconvenient time.)

Source: Cougar and McFadden, *Introduction to Computer Based Information Systems.* (New York: John Wiley & Sons, 1975): page 308.

readability generally should never be sacrificed in the name of efficiency. Contrary to the beliefs of many programmers, most programs run more efficiently if they are written with simple logic.[8] Computational "tricks" designed to improve execution speed often trick the user and greatly complicate future maintenance.

Before any change in a program is attempted, it should be benchmarked and measured for efficiency. If a bad logic segment or algorithm is found, the programmer will usually find it better to rewrite the code affected.

The tribulations of system and program design are somewhat facetiously summarized by the Laws of Programming in Table 7.1.

Programming style may be reduced to the following primitive elements:

1. Meaningful variable names
2. Clear, structured logic (GOTO-less)
3. Single-function subroutines
4. Use of comments at beginning of subroutines to describe inputs/outputs and function
5. Limited use of internal comments; if the code requires comments, rewrite the code so it is self-documenting
6. Modules limited to 50–100 lines of code

DEBUGGING

Recent studies have shown that the number and complexity of bugs can be greatly reduced if the program is designed using a top-down, modular approach.[16] As we saw in chapter 3, top-down design is an approach which takes a functional viewpoint. This process insures that all interfaces, logic decisions, and data structures are known before the coding begins.[6] Such an approach avoids simultaneous, inconsistent interfacing and reduces the complexity of the design. It also provides for orderly logic development. HIPO (tree structure) diagrams and pseudo-coding are two basic tools used in this type of design.

Structured design focuses on the functions to be performed. The emphasis is not on when or how many times the function is to be performed, but on what the inputs and outputs to the function are. There are three types of structured blocks or function modules used in structured design. They are:

Source—a block which gives and does not get

Sink—a block which gets but does not give

Transform—a block which gives and gets

As in the HIPO technique, the basic logic design problem is attacked first and then later each logical component is further broken down.

When either HIPOs or structured flowcharts are used to develop the logic flow of a program, they begin with a general module and decompose it until the most elementary logic modules are found. Thus the key idea behind top-down design is that if the program can be broken down into its smallest subfunctions, each subfunction can be coded separately and treated as a separate program. The greatest advantage here is that the resulting logic will be more coherent and that each subfunction of a program can be coded and recoded without significantly changing large segments at higher levels of the program design hierarchy. Procedural and block-structured languages facilitate this implementation process.

The following list will be useful when developing and improving program readability and accuracy:

1. Develop accurate tree structure reflecting program logic.
2. Number source logic segments for cross reference to design documentation.
3. Maintain source segment lengths at 100 statements maximum, excluding variable specifications.
4. Use variable names which reflect their function in the program.
5. Arrange procedures or segments within the source listing such that they appear in ascending numerical order, thereby allowing particular sets of logic to be retrieved easily.
6. Use the PL/I CALL or COBOL PERFORM command to implement at least the top logic structure. Lower levels can be set up in a similar fashion or with in-line code, depending upon complexity of program.
7. Each procedure or paragraph should have a brief comment inserted in the source code outlining its function.
8. Structured programming code structures are rigidly adhered to, except for error-handling segments.
9. Use spacing and indentation of source logic to highlight logical structure.

Common Program Development Errors

There are many classifications given to errors encountered during the debugging phase of program design. This discussion of errors will begin with a general categorization of error sources followed by further categorizations of program(mer)-attributed errors. Brown[2] and Sherman[13] provide a consensus categorization of system development error sources with an approximation of their relative frequency as follows:

Error Type	Estimated Frequency (%)
Program(mer) error	90
Operator error	5
Systems error	2
Software error	2
Hardware error	1

Program(mer) error is the largest category of system error. It is also the point of focus for this discussion. There are three major categories of program(mer) errors. They are:

Errors made outside the program

Errors made during programming

Errors in computation

Errors made outside the program. This type of error is made even before the design phase begins. The usual mistakes are faulty definition of the problem to be solved or poor formulation of the problem.[2] *Algorithm errors* are those made as a result of selecting a method, approach, or algorithm which is not suited to the problem.[13] *Analysis errors* result from formulation which does not focus with sufficient logical detail upon the various functions of the program.

Common analysis errors include:

1. Improper linkages and interfaces between modules
2. Improper sharing of variables among many modules
3. Failure to anticipate exceptional I/O conditions
4. Improper handling of exceptional cases (e.g., overflow, zero divide, etc.)[16]

Care must be taken to resolve and repair algorithmic- and analysis-type errors both during early program design and later as the program is being developed.

Errors made during programming. Debugging is the process of locating and correcting errors in a program. It is typically the most frustrating and time-consuming part of creating a working program. Generally, programming errors are due to oversights on the part of the programmer, resulting in translation problems by the compiler. Even the best programmers make mistakes. There are no cut and dried rules which, if followed, will guarantee an error-free program. However, many errors can be avoided by a logical top-down approach to program creation, making a careful check of all data input to the program, and then testing the program at each segment to evaluate what is actually generated.

Where possible, initial test runs should be made using a checkout or debugging compiler. This will aid in error identification, and upon encountering errors, these compilers will attempt to continue translation. The following seven error types dominate the early debugging stages:

1. Violation of basic syntax rules regarding semicolons, commas, parentheses or key words.
2. Misspelled variable name or key word.
3. Comment improperly formed.
4. Improper construction of logic block organization causing catastrophic control branching problems (e.g., PL/I END statements and COBOL paragraph termination).
5. Literal string not properly closed, thereby converting the remainder of the program into a literal string (try omitting the closing quote on a literal string to fully appreciate this one).
6. Illegal use of areas outside the source margin.
7. Improper or missing initialization of variables and constants.

This small checklist should be reviewed before attempting to find more subtle programming errors. Debugging is frequently a process of sequential removal of errors.

There are countless ways to abort the compilation of a program. It is, therefore, virtually impossible to create a compiler that can handle any error a programmer may make. Instead, thousands of coded error messages have been created, so whenever an error occurs a diagnostic message is printed. Generally, there are four levels of diagnostic messages that may occur during the compilation and execution phases. These are summarized in Table 7.2.

When interpreting diagnostic messages, the programmer must be aware that an error in one statement may cause several subsequent statements to be flagged as being in error. Yet, there may actually be nothing wrong with them. When the real statement in error is corrected, all subsequent related errors will automatically clear up.

Diagnostics are generated at two different times: during compila-

Table 7.2. Error Categories

Category	Description
(1) Warning Message	Calls attention to a possible error. Although the statement to which it refers is syntactically valid. The compiler has been designed to expect certain omissions in a program and these omissions are not really considered errors. The error is identified, then corrective measures are taken, pointing out what is missing. Complication continues with the assumptions indicated.
(2) Error Message	Signals an attempt to correct an erroneous statement, and the programmer is informed of the correction. Errors do not normally terminate processing of the source text.
(3) Severe Error	Indicates an error which cannot be corrected by the processor. The incorrect section of the program is deleted, but compilation is continued. Where reasonable, the error condition will be raised to mark the object module nonexecutable if execution of an incorrect source statement is attempted.
(4) Termination Error	An error so severe and unresolvable that it forces termination of compilation.

tion and while the program is executing. When detected during compilation of the program, they are known as compile-time errors. The diagnostic messages for compile-time errors are printed after the source listing (except in some debugging compilers where they are imbedded in the printout of the source). Errors detected during execution are known as execution or object-time errors and they are printed with the output of the program. General explanation of diagnostic messages can be found in accompanying literature which documents the particular compiler being used.

Compile-time errors are generally more numerous than object-time errors. Such items as keypunching and clerical errors generally fall into the compile-time category, although they may survive compilation to cause an error during execution. One way to avoid such errors is to make a careful manual inspection (desk checking) of the source input before submitting it for compilation. However, this is a very tedious job and there is still a possibility that all such errors will not be caught.

Object-time errors are less numerous than compile-time errors, but they can be much harder to find and correct. They are often a result of poor design or oversight of the programmer and are typically logical or structural in nature. Once again, the use of a debugging compiler can sometimes help the programmer find such errors, but all compilers are less effective against object-time errors than against compile-time errors.

Errors in computation. The programmer must be aware when and where computation errors may occur during program execution. Major categories of computation errors are as follows:

1. *Measurement errors* occur when digital numbers are used to try to quantify a value that cannot be measured precisely.
2. *Formulation errors* occur when approximations are used in problem-solving techniques.
3. *Truncation errors* occur when the number of significant terms of a value exceeds the finite number held by the computer.
4. *Rounding errors* occur when the size of an output value is limited by the size of the output field.

Knowing that many different types of errors occur, the programmer should try to minimize their effects and eliminate them through careful design and testing. It is not feasible to test all possible error conditions in a large program and, therefore, a somewhat arbitrary decision must be made as to when a program has been tested enough. Van Tassel suggests the following minimum number of tests be run:[14]

each module	4 runs
complete program	4 runs
live data	3 runs
system test	2 runs

Unfortunately, quantity of testing does not insure correctness. Other factors to consider are:

1. How important is arithmetic accuracy?
2. How often is the program to be used?
3. Over how long a period will it be used?

Program Style and Debugging

Table 7.3. Common Programming Errors (prior to execution)

A. Syntax Error Sources	B. Key Word Errors
1. Source margin error	1. Misspelled key word
2. Missing or unbalanced parentheses	2. Omitted key word
3. Extra or missing commas	D. Job Control Language Errors
4. Failure to properly terminate a literal string	1. Omitting or misspelling key words
5. Omit required blank spaces	2. Omission of required blanks
6. Failure to construct comment statements properly	3. Error in instruction sequence
	4. Failure to identify data set properly
C. Variable Specification Errors	5. Omission of JOB terminator
1. Failure to specify variable properly	6. Illegal use of column one by source or data segments
2. Improper constant specification	
3. Failure to initialize or reset value	
4. Failure to define constant value	

The data to be used in testing should include randomly selected general values from actual data, as well as extreme data values which test the full range of variables, subscripts, and iterative loops. All reasonably likely user errors should be included in the test data set.

As a logic error is detected and repaired during testing, the program must be run again to check for newly produced errors. Each time a program is modified, it must be retested with the old data, or the same data if possible, to check for different answers.

DEBUGGING CHECKLIST

Previous experience in training programmers has uncovered the following five major categories of errors:

1. Syntax
2. Key word usage
3. Declaration and initialization of variables
4. Job control language
5. Execution time errors

Table 7.3 summarizes common programming errors made prior to

Table 7.4. Common Execution Time Errors

---

1. Inaccessible statements in program
2. Improper termination of program statements
3. Errors in nested IF statements
4. Errors in constructing compound boolean expressions
5. Failure to properly terminate DO groups and procedures
6. Errors in complicated computational sequences
7. Addressing problems (array and string)
8. Improper data formats (reads and writes)
9. Off-by-one logic errors
10. Problem of first-card processing
11. Problem of last-card processing
12. Variable arithmetic overflows and underflows
13. Incorrect entrance into or exit from subroutines
14. Failure to evaluate operating system return and completion codes before continuation of processing

---

execution. Table 7.4 outlines typical problems encountered during execution. Of course, no simple checklist will find specific errors made. The purpose here is to highlight major areas to watch out for in hope of avoiding their consequence. Program debugging should be approached in a preventative mode, rather than a defensive one.

CONCLUSION

The subject of debugging is difficult to discuss abstractly and its skills are learned through experience. However, an attempt has been made here to call attention to common programming problems, particularly related to PL/I and COBOL, plus introduce certain checklists and techniques to aid in diagnosis. Each compiler has different approaches to syntax scanning and error correction. Some of these variations are of minor significance, while others can be quite confusing. It is generally recommended that large programs be segmented into manageable size and all initial debugging runs be made using a checkout or debugging compiler where the intent is to aid the programmer. As the modules (subsets) are debugged an optimizing compiler can then be used to create high-quality production modules. It is the programmer's responsibility to verify his/her logic design. This can be made more difficult by poor design, lack of documentation,

Program Style and Debugging

and sloppy code (constructs, variable names, indentation, style, etc.). It is hoped that some of the issues raised here will serve to bring attention to this aspect of programming.

DEBUGGING CASE STUDY

Debugging techniques may be learned, however it is during the application of these techniques that the theory becomes a working art. The example discussed here is designed to illustrate realistically the process of removing errors from a program. By slowly analyzing each section of code and the related logic, an attempt is made to simulate the sequential nature of the process. Every effort is made to involve the reader in these steps and a final debugged program appears in the appendix. A note of caution: even though the desired output has been generated, the overall style of the final program is far from perfect and may in fact still contain minor errors. Possibly even this will indicate how programs get loaded into a production environment too early.

The program similar to the one shown in Figure 7.1 was actually published in a national magazine as a working example of a biorhythm program. In fairness, it was originally written in BASIC and translated into PL/I (it didn't work in BASIC either). The theory of biorhythms is that we all have three metabolic cycles: The physical cycle (23 days), the emotional cycle (28 days), and the intellectual cycle (33 days). All of these cycles begin at birth, moving upward, at the point of birth, and continue throughout our lifetime. Our well-being then is reflected as the relative position of each curve. (Interested readers can refer to numerous "technical" sources on the subject.) The real concern here is why the existing program does not work.

Basically, the program in Figure 7.1 is designed to determine the number of days since birth, then compute the following curves:

P = Physical
I = Intellectual
E = Emotional
A = Average of P, I and E

Figure 7.2 illustrates the Biorhythm Chart for the first 40 days of an individual's life where A equals the average for that particular day. Specific data for this program are as follows:

1. Name. Example 'Barbara Stockwell'

Figure 7.1: Biorhythm plotting program—Version I

```
STMT  LEV NEST                SOURCE PROGRAM TEXT

               % /*    BIORHYTHM PLOTTING PROGRAM VERSION I    */
  1    1       % BIO: PROCEDURE OPTIONS (MAIN);
  2    1       % DECLARE      T$ (8)                 CHAR (3)
               %              INIT ('SUN', 'MON', 'TUE', 'WED', 'THU', 'FRI', 'SAT', '   '),
               %              K                      FLOAT (4) INIT (6.283185),
               %              F (12)                 FIXED
               %              INIT (31, 28, 31, 30, 31, 30, 31, 31, 30, 31, 30, 31),
               %              Z$                     CHAR (20),
               %              A (51)                 CHAR (1),
               %             (J1, J2)                FLOAT (8),
               %              L                      FLOAT (8),
               %             END_FLAG                BIT (1) INIT ('0'B);
  3    1       %   ON ENDFILE (SYSIN) END_FLAG = '1'B;
  4    1       %   A = ' ';
               %   /* DATA INPUT SECTION */
  5    1       %   GET LIST (Z$);
  6    1       % LOOP:
  7    1    1  %   DO WHILE (END_FLAG = '0'B);
  8    1    1  %        PUT SKIP LIST ('ENTER DATES IN THE FORMAT MM,DD,YYYY');
  9    1    1  %        PUT SKIP LIST ('EXAMPLE-JUNE 16,1944 WOULD BE 6, 16, 1944');
 10    1    1  %        PUT SKIP LIST ('  ');
 11    1    1  %        PUT EDIT ('BIRTH DATE') (SKIP, COL (11),A);
 12    1    1  %        GET LIST (M1, D1, Y1);
 13    1    1  %        IF Y1 > 99          THEN            GO TO L255;
 14    1    1  %        Y1 = Y1 + 1900;
 15    1    1  %        PUT EDIT (Y1) (COL(25),F(4,0));
 16    1    1  % L255:
 18    1    1  %        M2 = M1;
 19    1    1  %        D2 = D1;
```

```
20  1                          Y2 = Y1;
21  1                          CALL DAY;
22  1                          P1 = N7;
23  1                          IF N7 > 7           THEN          P1 = 8;
24  1                          PUT SKIP LIST ('START DATE FOR CHART');
25  1                          GET LIST (M2, D2, Y2);
26  1                          IF Y2 > 99          THEN          GO TO L295;
27  1                          Y2 = Y2 + 1900;
28  1                          PUT EDIT (Y2) (COL (25), F(4,0));
29  1           L295:          CALL DAY;
31  1                          P2 = N7;
32  1                          IF N7 > 7           THEN          P2 = 8;
33  1                          PUT SKIP LIST ('LENGTH OF CHART IN DAYS');
34  1                          GET LIST (L);
35  1           /*  CALCULATE OFFSET, TAKING LEAP YEARS INTO ACCOUNT    */
36  1                          X = M1;
37  1                          CALL JULIAN;
38  1                          J1 = J2 + Y1 * 365;
39  1                          IF J1 < 639723      THEN          P1 = 8;
40  1                          X = M2;
41  1                          CALL JULIAN;
42  1                          J2 = J2 + Y2 * 365;
43  1                          IF J2 < 639723      THEN          P2 = 8;
44  1                          OH = J2 - J1 + 4 * FLOOR ((Y2 - Y1) / 4 - FLOOR ((Y2 - Y1) / 4));
45  1                          IF Y1 / 4 - FLOOR (Y1 / 4) = 0    THEN
46  1                                                                          GO TO L400;
47  1                          IF Y2 > Y1                        THEN          GO TO L370;
48  1                          IF M2 > 2                         THEN          GO TO L370;
49  1                          GO TO L400;
50  1           L370:
```

```
STMT  LEV NEST              SOURCE PROGRAM TEXT

 52    1   1                   IF M1 < 3          THEN        OH = OH + 1;
 53    1   1   L400:
                /*  PRINT CHART HEADER      */
 54    1   2           DO I = 1 TO 5;
 55    1   2              PUT SKIP LIST (' ');
 56    1   1           END;
 57    1   1           PUT EDIT ('BIORHYTHM CHART FOR ',Z$)
 58    1   1                    (SKIP, COL(20),A);
 59    1   1           PUT SKIP LIST (' ');
 60    1   1           PUT EDIT
                            ('BORN ON ', T$(P1),' ',M2,'/',D1,'/',Y1)
 51    1   1                  (SKIP, COL(28),3A,F(2,0),A,F(4,0));
 62    1   1           PUT EDIT
                            ('BEGINNING', T$(P2),' ',M2,'/',D2,'/',Y2)
 63    1   1                  (SKIP, COL(28),3A,F(2,0),A,F(4,0));
 64    1   1           PUT SKIP LIST (' ');
 65    1   1           PUT EDIT
 66    1   1           PUT EDIT
                            ('P = PHYSICAL           (23 DAYS)')  (SKIP,COL (28),A);
 67    1   1           PUT EDIT
                            ('E = EMOTIONAL          (28 DAYS)')  (COL (28),A);
 68    1   1
 69    1   1           PUT EDIT
 70    1   1                ('I = INTELLECTUAL      (33 DAYS)')  (COL(28),A);
 71    1   1           PUT EDIT
 72    1   1                ('A = OVERALL AVERAGE')  (COL(28),A);
 73    1   1           PUT SKIP LIST (' ');
 74    1   1           PUT EDIT ('DOWN','CRITICAL','UP')
 75    1   1                  (SKIP, COL(13),A,COL(34),A,COL(62),A);
 76    1   1           PUT EDIT ('----------------------------------------------')
 77    1   1                  (COL (13), A);
```

162

```
 78  1       /*  SET F (2) TO 29 FOR LEAP YEARS    */
 79  1           IF Y2 / 4 - FLOOR (Y2 / 4) = 0
     1              THEN       F (2) = 29;
 80  1       /* GENERATE THE BIORHYTHM PLOT        */
 81  1   2      L = OH + L;
 82  1   2      C = 0;
 83  1   2      DO OH = OH TO (L-1);
 84  1   2         OH = OH +1;
 85  1   2         C = C + 1;
 86  1   2         Y = 0;
 87  1   2         X = (SIN (K * (OH / 23 - FLOOR (OH / 23))) * 25) + 26;
 88  1   2         A (X) = 'P';
 89  1   2         Y = Y + X;
 90  1   2         X = (SIN (K * (OH / 33 - FLOOR (OH / 33))) * 25) + 26;
 91  1   2         A (X) = 'I';
 92  1   2         Y = Y + X;
 93  1   2         X = (SIN (K * (OH/28 - FLOOR (OH/28))) * 25) + 26;
 94  1   2         A (X) = 'E';
 95  1   2         Y = (Y + X)/3;
 96  1   2         A (Y) = 'A';
 97  1   2         PUT EDIT
     1   2            (T$(P2),M2,'/',D2,A)
     1   2            (SKIP,A,COL(5),F(2,0),A,F(2,0),COL(13),51A);
     1       /* INCREMENT DATE                      */
 98  1   2         IF P2 = 8       THEN      P2 = 1;
 99  1   2         P2 = P2 + 1;
100  1   2         IF P2 > 7               THEN       P2 = 1;
101  1   2         D2 = D2 + 1;
102  1   2         IF D2 > F (M2)          THEN DO;
104  1   3            D2 = 1;
105  1   3            M2 = M2 + 1;
```

163

```
STMT  LEV NEST              SOURCE PROGRAM TEXT

106    1    3               END;
107    1    2               IF M2 < 13          THEN      GO TO L640;
108    1    2               M2 = 1;
109    1    2               Y2 = Y2 + 1;
110    1    2     L640:
112    1    2               IF Y2/4 - FLOOR (Y2/4) ¬= 0   THEN   GO TO L655;
113    1    2               F (2) = 29;
114    1    2               GO TO L660;
115    1    2     L655:
117    1    2               F (2) = 28;
118    1    2     L660:
120    1    1               END;
121    1    1               GET LIST (Z$);
122    1    1             END LOOP;
                 /*  FIND DAY OF THE WEEK    */
                 DAY:        PROC;
123    2    2               N1 = M2 + 12 * FLOOR (.6 + 1 / M2);
124    2    2
```

```
125  2        N2 = Y2 - FLOOR (.6 + 1 / M2);
126  2        N3 = FLOOR (13 * (N1 + 1) / 5);
127  2        N4 = FLOOR (5 * N2 / 4);
128  2        N5 = FLOOR (N2 / 100);
129  2        N6 = FLOOR (N2 / 400);
130  2        N7 = N3 + N4 - N5 + N6 + D2 - 1;
131  1     END;
           /* FIND DAYS EXPANDED IN PRIOR MONTHS      */
   JULIAN:   PROC;
132  2        J2 = 0;
133  2        DO I = 1 TO (X - 1);
134  2   1       J2 = J2 + F (I);
135  2   1    END;
136  2     END;
137  1  END BIO;
138     'BARBARA STOCKWELL'
        07,11,1955
        07,11,1955
        40
```

Figure 7.2: Hypothetical output of biorhythm program

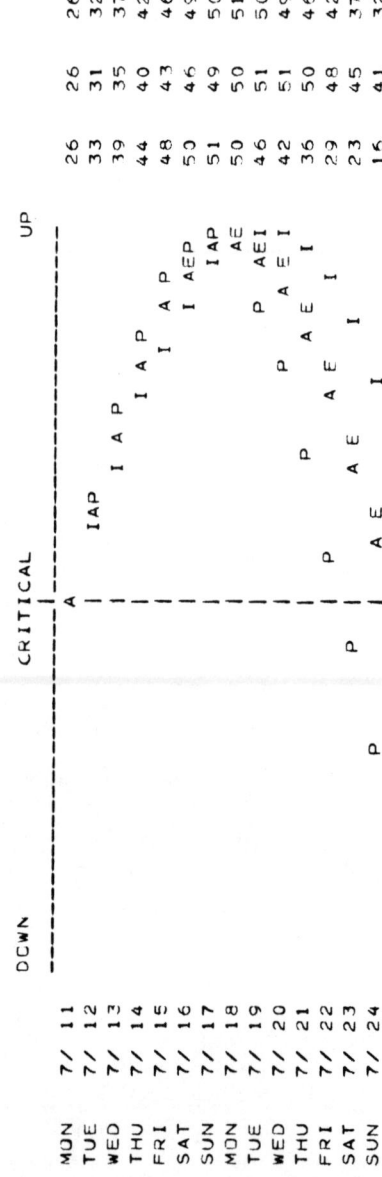

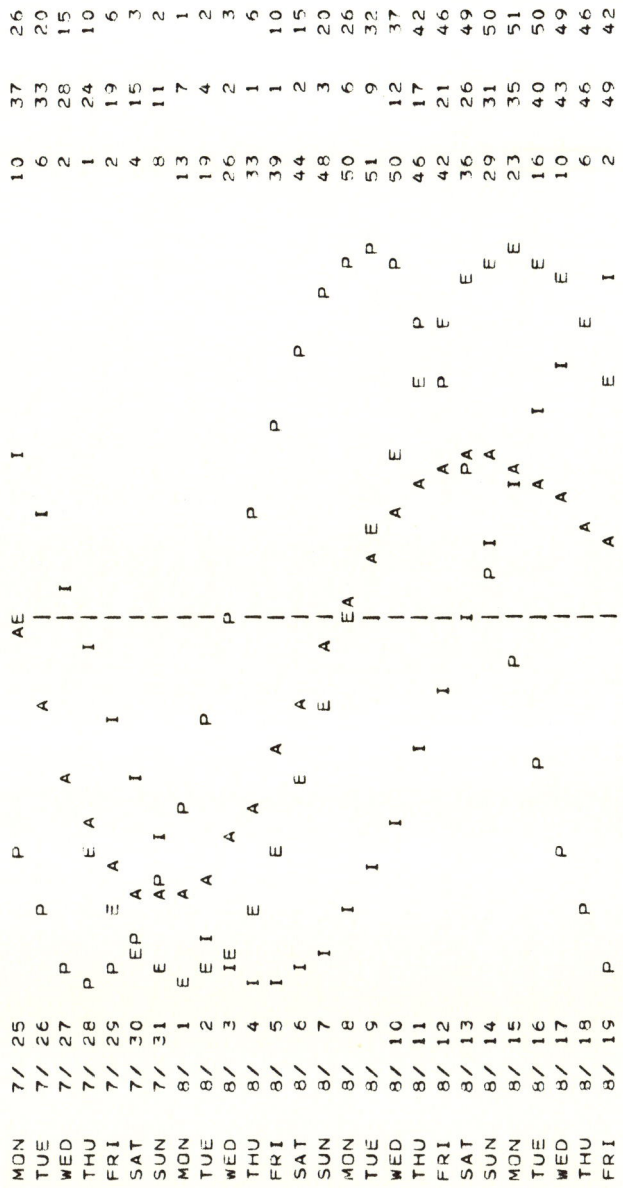

Figure 7.3: Actual output of biorhythm program (produced from program shown in Figure 7.1)

```
ENTER DATES IN THE FORMAT MM,DD,YYYY
EXAMPLE-JUNE 16, 1944 WOULD BE 6, 16, 1944
          BIRTH DATE
START DATE FOR CHART
LENGTH OF CHART IN DAYS



          BIORHYTHM CHARTER FOR
          BARBARA STOCKWELL

               BORN ON        7/11/1955
               BEGINNING      7/11/1955

               P=PHYSICAL            (23DAYS)
               E=EMOTIONAL           (28DAYS)
               I=INTELLECTUAL        (33DAYS)
               A=OVERALL AVERAGE


          DOWN           CRITICAL              UP
          ------------------|------------------
```

MON 7/11										
TUE 7/12										
WED 7/13										
THU 7/14										
FRI 7/15							IAP			
SAT 7/16							IAP			
SUN 7/17							IAP			
MON 7/18							IAP	I A P		
TUE 7/19	P						IAP	I A P		
WED 7/20	P E P A			AE I			PIAP	I A P	I A P	
THU 7/21	E U E P A		EA	AE I			AIEP	A L P	I A P	APA
FRI 7/22	EIE P A	AA PI	P	AE I			AIEPI	A I P	I A P	IAEPI
SAT 7/23	IIE P A	EA PI	P	AE I			AIEPI	A I P	I A P	IIEPI
SUN 7/24	III P A	EA PI	A	AE A			AIEPI	A I P	I A P	IIEPI
MON 7/25	III P AI	EA PI	A	AE A			AIEPI	A I P	I A P	IIEPI
TUE 7/26	III P AI	EA PI	I	AE A	P		AAERI	A I P	I A P	LIEPI
WED 7/27	III P AI	EA PI	I	AE A	P		AAEPIA	A I P	I A P	IIEPI
THU 7/28	III P AI	EA PI	I	AE PA	P		AAEPIAPA	PA E P	I A P	IIEPI
FRI 7/29	III P AI	PA PI	I	AE PA	I		AAEPIAIA	PA E P	I P P	IIPPI
SAT 7/30	III P AI	PA PI	I	AE PA	I		AAEPAAIA	PA E P	I P P	IIPPP
SUN 7/31	III P AI	PA PI	I	AE PA	I		AAEPAAIA	PA E P	I P P	PIPPP
MON 8/1	III P AI	PA PI	I	AE PA	I		AAEPAAIA	PA E P	I P P	PIEPP
TUE 8/2	III P AI	PA PI	I	AE PA	I		AAEPAAIA	PA E IP	I P P	PIEPE
WED 8/3	IPI P AI	PA PI	I	AE PA	L		AAEPAAIA	PA E LP	I P P	PIIPE

169

2. Birth date (MM,DD,YYYY). Example 7,11,1955
3. Start date for chart (MM,DD,YYYY). Example 7,11,1955
4. Number of day to print output. Example 40

The output generated using this data and the program are presented in Figure 7.3. Note that the printed results indicate severe problems exist with the present logic.

After spending some time analyzing the program, the chart length error appears to lie within the DO loop. Note that OH is the variable being incremented in each iteration. However, in statement 83, OH is again incremented causing it to increase by two each iteration, rather than one. This error may be corrected by removing the incrementing instruction 83.

The chart printing error also appears to be in the DO loop. After the data for each day have been plotted, the designated output area should be cleared. The area is cleared originally by statement 4, however this occurred only at the beginning of the program. Therefore, all previously plotted cycle values are retained and printed for each iteration. To correct this error, statement 4 should be moved to the beginning of the DO loop.

A third major error is contained in the DAY procedure. When calculated correctly, N7 should represent the day of the week. For individuals born on July 11, 1955, the values calculated in subroutine DAY should be shown in Figure 7.4.

P1 and P2 receive the computed N7 value in statements 22 and 32. These values then become the subscript for T$ in statements 60, 62, and 97. Because P1 and P2 must have a value between 1 and 8, the author inserted IF statements to test for this condition (statements 23 and 33). The values of P1 and P2 were both greater than 7 so they were assigned the value of 8. T$(8) corresponds to a blank space.

This error has not yet been corrected. After some analysis the computational algorithm for N7 was revised. The new equation is as follows:

$$N7 = MOD(N7,7) + 1$$

The proper value for N7 is now calculated as

$N7 = MOD(N7,7) + 1 = MOD(2458,7) + 1 = 2$ (e.g., signifies Monday)

By using this corrected value of N7, P1 and P2 now will contain the proper values to be subscripts of T$. The day designated as T$(2) is

Figure 7.4: Manual calculations of DAY procedure key values

N1=M2+12*FLOOR(.6+1/M2)=7+12*FLOOR(.6+1/7)=7
N2=Y2-FLOOR(.6+1/M2)=1955-FLOOR(.6+1/M2)=1955
N3=FLOOR(13*(N1+1)/5)=FLOOR(13*(7+1)/5)=20
N4=FLOOR(5*N2/4)=FLOOR(5*1955/4)=2443
N5=FLOOR(N2/100)=FLOOR(1955/100)=19
N6=FLOOR(N2/400)=FLOOR(1955/400)=4
N7=N3+N4−N5+N6+02-1=20+2443-19+4+11-1=2458

Monday, which is the correct day of the week for the date in question.

Since the major logical errors have been corrected, program style issues can now be analyzed. The first area of questionable style lies in the use of GOTO statements. A well-structured program should *not* make free use of these statements. This program has eight GOTO statements. The segments of code dealing with these statements should be rewritten to produce a smoother flowing program. Elimination of GOTO statements will also reduce the need for so many labels, although they can remain to aid readability.

A program should be well structured and include enough internal comments to aid the reader. This program contains nine comment statements, and only one of the eight END statements has comments attached to indicate what blocks they end. Because this program has now been reformatted, each END statement has been aligned with its respective key word. However, additional comments added would further improve readability of the overall program. Comments prior to each subroutine describing the inputs, outputs, and function would also be a desirable addition.

The third area of style criticism is in the selection of variable names. Variable names and internal procedure names should attempt to represent accurately their purpose. In this program, only single-letter variable names have been assigned. The meaning of each variable must, therefore, be gathered by studying the entire program. As an example of proper naming, the internal procedure names Day and Julian appropriately identify their function.

Once these style changes have been made, other individuals, besides the author, should be able to comprehend the purpose and function of the program. Of course, good accompanying logical documentation such as tree diagrams and pseudo-code would also help this process.

At this point the reader must spend considerable time with this example to get full value from it. The program should be executed and errors removed until the biorhythm output is correct. Select appropriate test data to check out the program. A logically correct solution is presented in the appendix. Examine this solution and answer the associated discussion questions.

Review Questions

7.1. What are some of the programming implications of using efficiency tricks in a program design?

7.2. In what aspect of the software development process do you anticipate greatest change over the next ten years? Discuss.

7.3. Define programming style.

7.4. To what degree should comments be used in program design? Why do some programmers differ on the answer to this question?

7.5. Illustrate the "off-by-one" logic error with skeleton code. How can you find logic errors of this type in a program during debugging?

7.6. What is the definition of a "working" program? Describe how you might proceed to develop an error-free program.

7.7 Outline some of the more common error types made during the early debugging process. Which one gives you personally the most trouble? Why?

7.8 What are the error types found during computation? Develop examples of each.

7.9. Summarize the common errors made during and prior to program execution.

Notes and References

1. Bates, Frank, and Douglas, Mary L. *Programming Language/One*, Third Edition. Englewood Cliffs, N.J.: Prentice-Hall, Inc., 1975.
2. Brown, A.R., and Sampson, W.A. *Program Debugging.* New York: American Elsevier Publishing Company, Inc., 1973.
3. Cougar, Daniel, and McFadden, Fred J. *Introduction to Computer Based Information Systems.* New York: John Wiley & Sons, 1975.
4. Dahl, O.J.; Dijkstra, E.W.; and Hoare, C.A.R. *Structured Programming.* New York: Academic Press, 1972.

5. Goos, G., and Hartmanis, J., eds. *Lecture Notes in Computer Science,* volume 23, "Programming Methodology." New York: Springer-Verlag, 1975.
6. Goos, G., and Hartmanis, J., eds, *Lecture Notes in Computer Science,* volume 36, "Theory of Program Structure, Schemes, Semantics, Verification." New York: Springer-Verlag, 1975.
7. Hughes, Joan K. *PL/I Programming.* New York: John Wiley & Sons, 1973.
8. Kernighan, Brian W. and Plauger, P.J. *The Elements of Programming Style.* New York: McGraw-Hill Book Co., 1974.
9. Ledgard, Henry F. "COBOL Under Control," *Communications of the ACM,* 19(11): pp. 601–608.
10. Richardson, Gary L., and Birkin, Stanley J. *Problem Solving Using PL/C.* New York: John Wiley & Sons, 1975.
11. Rustin, Randall, ed. *Debugging Techniques in Large Systems.* Englewood Cliffs, N.J.: Prentice-Hall, Inc., 1970.
12. Sellers, Frederick F., Jr.; Hsiao, Mu-Yue; and Bearnson, Leroy W. *Error Detecting Logic for Digital Computers.* New York: McGraw-Hill Book Co., 1968.
13. Sherman, Philip M. *Techniques in Computer Programming.* Englewood Cliffs, N.J.: Prentice-Hall, Inc., 1970.
14. Van Tassel, Dennie. *Program Style, Design, Efficiency, Debugging, and Testing.* Englewood Cliffs, N.J.: Prentice-Hall, Inc., 1974.
15. Weinberg, Gerald M. *The Psychology of Computer Programming.* New York: Van Nostrand Reinhold Company, 1971.
16. Yourdon, Edward. *Techniques of Program Structure and Design.* Englewood Cliffs, N.J.: Prentice-Hall, Inc., 1975.

8
DATA BASE CONSIDERATIONS

Program design has been viewed thus far as essentially an internal structuring process. Actually there are several data-related considerations which can affect the design process. This chapter focuses on the relationships of the data base structure to program design.

Data base—the very mention of the term conjures up images of vast, nonredundant aggregations of data, located just a disk drive away. Theoretically, anyone who needs information can quickly retrieve it from the data base.

But wait a minute, what purpose does a discussion on data base considerations serve in a primer on structured program design? A data base's structure affects program design. From the simplest to the most complex system, program design will be affected by the content and organization of the data base. Peters and Tripp[2] in their comparison of five currently popular program design techniques encountered logical design flaws when using a "data base" that was alien to a specific program design technique. Jackson[1] describes the same point with his concept of structure clashes. Even the novice programmer recognizes the rudimentary elements of the problem whenever he/she first becomes aware that the data must be "reorganized" (or sorted) to solve a particular data manipulation problem.

Almost every program requires a data base and the scope of that data base includes all of the internal variables used by the program, plus the type and structure of any external variables used. Usually the term "data base" refers to commercially oriented processing systems that manage large collections of external data as an integrated whole. This software technology is designed to reduce the artificial boundaries imposed by separate files, plus it permits users to access data in a more "natural" manner. However, a complete description of today's

A Primer on Structured Program Design

Figure 8.1: Selection of variable types

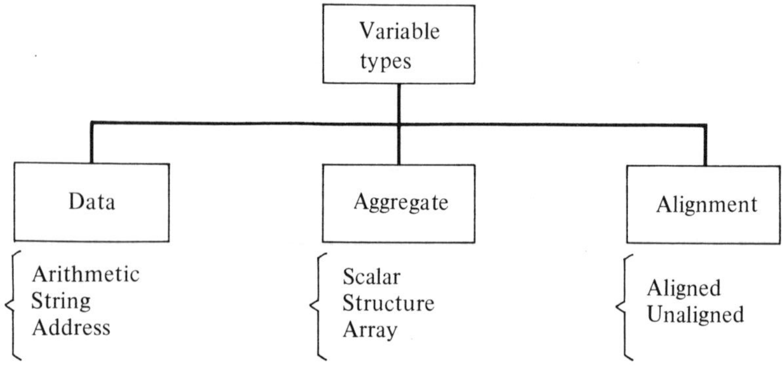

```
                    ┌──────────┐
                    │ Variable │
                    │  types   │
                    └────┬─────┘
         ┌───────────────┼───────────────┐
    ┌────┴───┐      ┌────┴────┐     ┌────┴────┐
    │  Data  │      │Aggregate│     │Alignment│
    └────────┘      └─────────┘     └─────────┘
  ⎧ Arithmetic    ⎧ Scalar          ⎧ Aligned
  ⎨ String        ⎨ Structure       ⎨ Unaligned
  ⎩ Address       ⎩ Array           ⎩
```

current data base technology and available software systems is beyond the scope of this discussion. Rather, this chapter's goal is to describe the primary elements to be considered in selecting and organizing the data used internally within a program. In addition, this chapter will briefly describe some external data structuring techniques that are commonly used in various programming applications.

INTERNAL DATA BASE

To a COBOL programmer accustomed to explicitly describing each and every variable in a separate data division, the wild abandon with which FORTRAN programmers create and use variables must be frightening. PL/I, with its default variable rules and block structure which dynamically create and reference variables, must be equally as alien to both the FORTRAN and COBOL programmer. Regardless of the host language, all data variable options are essentially similar. A principal requirement for efficient execution of any program requires a restriction on the kinds of values that can be assigned to a storage location. This restriction is generally achieved by associating specific storage types or attributes with each variable. Typically, variable specifications involve the following kinds of information:

1. The *data type* which specifies implicitly or explicitly the range and representation of storage for a simple datum, such as a number, a character string, a label, etc.
2. The *aggregate type* defining the way storage is arranged for

Figure 8.2:   Arithmetic variable choice

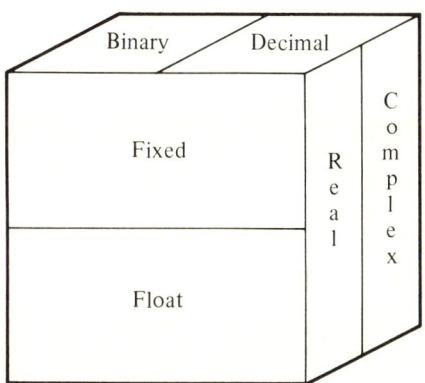

individual elements, arrays, or structures.
3. The *alignment type* indicating the specific manner in which storage is laid out in hardware memory which in turn determines memory and access requirements.

All programming languages allow for these variable types. COBOL and FORTRAN provide many of these different types as features of the language. PL/I allows the programmer complete freedom to specify any of the variable types desired or to optionally allow the internal language default rules to provide any omitted type(s). Figure 8.1 depicts the basic options of variable types available to the PL/I programmer.

Arithmetic Storage

Figure 8.2 illustrates the eight different choices available for representing arithmetic data (ignoring differences in scaling and precision). The selection of one of these eight choices is at least a compromise between program logic and hardware efficiency; for example, it may be more convenient to use decimal numbers, but binary numbers lend themselves to more rapid cpu processing. Figure 8.3 shows a small PL/I program which demonstrates the effect of storage mode upon processing speed. Here the program allocates various variable types, then performs a series of arithmetic operations repetitively. The results produced by executing this program and logically equivalent FORTRAN and COBOL versions are shown in Figure 8.4. This

Figure 8.3. PL/I Arithmetic Test Program

ARITH: PROCEDURE OPTIONS (MAIN);
  DCL (A,B,C,D,E,I,N) [variable specifications]
  GET LIST (N);
  DO I = 1 TO N;
    X = (A + B − C / D * E);
  END;
END ARITH;

---

graphically illustrates the effects of different variable data types on the same program. In addition, it provides a crude comparison between COBOL, FORTRAN, and PL/I on the identical program logic.

Arithmetic storage is used in practically all programs and programming languages. Table 8.1 illustrates the eight different PL/I arithmetic storage types with their common FORTRAN and COBOL equivalents. The choice of data types for numeric data elements is a major decision in program design. The primary choice is between the arithmetic storage types, described above, and the "pictured" storage type described later. Arithmetic storage types can be manipulated by the cpu more efficiently, but they also require conversion when input/output is required. Typically, arithmetic storage is selected over "pictured" storage whenever calculations are complex or when storage space utilization is critical. Once the arithmetic type has been selected for a variable, the additional attributes of scale, base, mode, and precision must also be selected. These selections are critical, since they affect the range of input data, accuracy of results, effort required in program development, and even the actual time of program execution.

Because each program has different processing requirements, each language different capabilities, each compiler different limitations, and each programmer different talents, no "universal" guidelines can be cited in selection of arithmetic storage attributes. Common sense, however, dictates that the *program's* requirements are the primary source of attribute selection.

Pictured String Storage

Pictured string storage provides an alternative to arithmetic and ordinary character string storage. Pictured storage is most often used

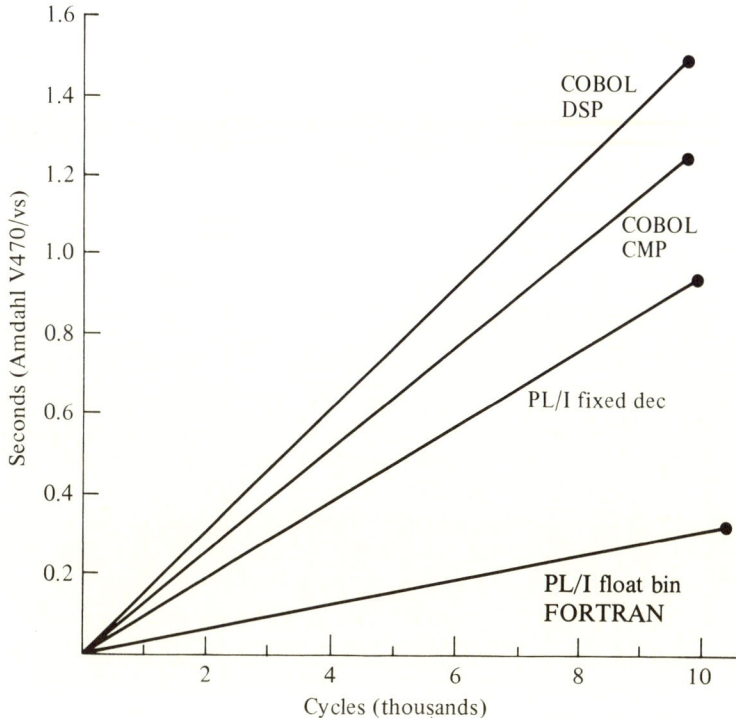

Figure 8.4: Variable type performance graph

in business applications and by COBOL programs because it eliminates the use of special representations for numeric values in the program. Many programs can be written in a highly natural style using pictured storage because it uses the same representation within the program

Table 8.1. Equivalent Arithmetic Storage Types

PL/I	FORTRAN	COBOL
Fixed Bin Real	Integer	Comp
Fixed Dec Real	——	Comp-3
Float Bin Real	Real	——
Float Dec Real	——	Comp-1

that human users do on reports, cards, and other ouput listings. Pictured storage is ideal for programming applications in which the formatting and layout of input/output is important and arithmetic calculations are relatively simple (e.g., addition and subtraction).

Unfortunately, when a pictured variable is involved in complicated and repeated arithmetic processing, a significant amount of processing overhead occurs. The retrieval and assignment times for arithmetic values to a pictured variable storage location can be several orders of magnitude greater than the retrieval and assignment times for a similar arithmetic variable. The impact of this overhead depends on the requirements of the program. The program designer has at least three options to deal with this overhead:

1. Accept the extra cost and keep the pictured variable. This is the approach adopted by COBOL where the input/output volume is assumed to overshadow additional computational costs incurred by the pictured storage.
2. Drop the use of pictured variables entirely and incur conversion overhead for all input/output. This is the approach used by FORTRAN and assumes only the most limited input and output.
3. Keep the pictured variable for input/output, but introduce an arithmetic variable for use in calculations. In this case the programmer explicitly controls and limits the overhead required by the pictured variables through assignment statements. This is the most efficient option if any degree of arithmetic is required since it retains the simplicity of input/output offered by pictured variables and expends the conversion overhead only once (e.g., on input or output).

Ordinary String Storage

String storage is divided into two major types: ordinary and pictured. The differences between the two types are more important than their similarities. Ordinary string storage is used primarily for text, while pictured (string) storage is normally (although not exclusively) used for numeric data that is internally represented as character strings.

Ordinary string storage data types have two attributes. These are the *string type* (characters or bits), and in both PL/I and COBOL the presence or absence of the *varying* attribute. The character string type is obviously needed for nonnumeric string character data. The bit string type consists of the binary characters "0" and "1", and is useful

Data Base Considerations

in representing boolean values (similar to FORTRAN LOGICAL or certain COBOL 88 level data variables).

Do not confuse bit strings with binary arithmetic storage, as bit variables are not *normally* used in calculations. Instead, bit strings provide a convenient and economical way to store binary status information (where "1" = TRUE and "0" = FALSE).

In both PL/I and COBOL, string storage without the varying attribute can accommodate a string of only one length—the maximum specified. Whenever a shorter string is assigned to a "nonvarying" data variable, padding occurs in the unfilled positions of the string. But PL/I and COBOL varying strings accommodate strings of any length (0 to the maximum) without padding. As might be expected, the varying attribute requires an additional storage overhead with each data variable to maintain the "current" length of the variable. Varying strings are therefore processed slightly less efficiently than nonvarying strings. Varying strings are typically used in application programs which require text manipulations on unformatted character data. (Note: Sometimes programmers are confused into thinking that the varying attribute "saves" storage space. This is *not* true. In fact, the opposite occurs, since the varying string *always* occupies the maximum storage space declared *plus* the overhead storage necessary to maintain the current string length.)

Address Storage

The address storage data type is not available to the COBOL and FORTRAN user, however both languages do use address storage data types whenever they "ALTER" a statement, or "FORMAT" a print line, or use any type of "GOTO." PL/I expands this capability by permitting an address variable to be treated like a data value. An address storage variable is not used in arithmetic calculations, however it can be retrieved, assigned, compared (with other address variables), and ultimately used as an address. Table 8.2 lists the six common address storage variables used in COBOL, FORTRAN, and PL/I.

Sometimes other key words are used to modify the attributes of an address variable. Note, however, that these additional key words are not part of the storage types but provide additional information about the usage of the address variables. For example, consider:

DECLARE TOM ENTRY;
DECLARE JERRY ENTRY (CHAR (*), FIXED BINARY, PTR);

Table 8.2. Address Types by Language

PL/I	FORTRAN	COBOL
label	statement no.	paragraph
entry	entry	entry
format	format	——
pointer	——	——
offset	——	index
file	file	file

Here the storage type of both TOM and JERRY is ENTRY, and the additional declarations following JERRY are about the usage of the address variables.

Aggregate Types

Many times it is useful to group together a set of individual or scalar variables, arrange them into some sequence or order, and then treat that sequence as a single, integrated variable. Variables treated in such a manner are called *aggregates*. Aggregates consist of two types: *structure* (heterogeneous) and *array* (homogeneous).

Within aggregates it is necessary to distinguish between levels of variables. A variable that is contained in another larger variable is a minor variable and is called a *component* (or member) of the aggregate. As would be expected, a variable *not* contained in another variable is called a *major variable* or *level one variable* (due to the way structures are defined).

Structures. Structure variables are heterogeneous sequences of members. A structure has a reference name, and each member of the structure has its own name. To reference the entire structure the level one variable name is used. To reference any member of the structure often requires *both* the structure name and its lower level member name to be used. In PL/I a fully qualified reference is separated by a period in the form:

<structure name> . <member name>.

The COBOL fully qualified reference uses the modifiers "of" or "in" to separate the structure and member names as follows:

<member name> OF <structure name>

Each component or major variable of a *structure* may have any storage type supported by the language. This is the *distinction* between structure and array types of aggregate storage. The first member could be a pictured string type. The second could be complex floating point, the third member an array of characters, and so on.

The structure is unique to both PL/I and COBOL. Structures are identified by level numbers preceding each name in the declaration of the structure. The level numbers for each member must be greater than the level number of the structure itself, since that is how the hierarchy is determined. For example, consider the following declaration of a structure:

```
DECLARE 1 STUFF,
    2 JUNK PICTURE '(4)X',
    2 TRASH,
        3 GARBAGE FIXED BINARY,
        3 WASTE FLOAT DECIMAL,
    2 SEWAGE (4) FLOAT; /* NOTE 4 ELEMENTS */
```

This specification illustrates how a structure member can be scalar, a structure, or an array. (The use of array here is premature, but the array STUFF SEWAGE is not overly complicated and its treatment should be obvious.) Notice also that the variables STUFF and TRASH do *not* have associated storage types. This omission of storage type and an intermediate reference level identifies these two variable-names as structures.

Structures are convenient methods for handling nonhomogeneous aggregates of members since the members can be handled collectively, or separately, according to the requirements of the program. Structures lend themselves to efficient input/output methods and for this reason are quite important in many commercial environments.

Arrays. An array is a sequence of elements each containing the same type of variable (e.g., a homogeneous list). They share a single name and the individual elements are designated by the array name and a subelement (subscript) identification. Within the program, general expressions can be used for the subscripts to allow an exact data element to be dynamically referenced during execution; hence a single array reference can designate different data elements at different times during program execution.

All components of an array have the same storage type. An array can be any storage type except another array. In PL/I arrays are declared by means of the dimension attribute which is normally placed directly after the name of the variable. As an example, let us use the following scalar and structure variables to illustrate these points further:

    DECLARE GAMMA FLOAT;

    DECLARE 01 DATE__REC,
       02 NAME CHAR (20),
       02 PHONE__NUMBER FIXED BINARY;

Each of these can be transformed into an *array* of *scalars* and an *array* of *structures* by the addition of an appropriate "dimension" attribute as follows:

    DECLARE GAMA (20,10) FLOAT BINARY:
    /* 20 ROWS & 10 COLUMNS */

    DECLARE 01 DATA__REC (30),
    /* 30 COPIES OF STRUCTURE */
       02 NAME CHAR (20),
       02 PHONE__NUMBER FIXED BINARY;

A FORTRAN programmer's fondness for arrays is probably surpassed only by a COBOL programmer's love of structures (records). However, since PL/I and COBOL use structures *and* arrays (separately and in combination), the decision on when to use either a structure or an array can be determined using the following guidelines:

1. The members of a structure can have different data types (character, binary, arithmetic, etc.), whereas the elements of an array must all have the same data type.
2. An element within an array can be selected and referenced at execution time via subscript value, but a member of a structure must be referenced explicitly at the source/code level.
3. Since arrays and structures can be combined in both PL/I and COBOL, it is possible to provide a reasonably good organization for a complicated data base.

Alignment

At the start of this section, it was indicated that each storage variable had a *data type* (arithmetic, string, address), an *aggregate type* (scalar, array, structure), and an *alignment type*. The first two

Data Base Considerations

determine what values can be stored in that storage variable, while the latter, alignment, affects the manner in which the variable is located in storage. The alignment specification is given as a single attribute in PL/I, either ALIGNED or UNALIGNED. Normally this attribute is omitted, since the compiler usually generates storage allocation for the most efficient cpu access. Unfortunately, this "efficient access" that the compiler so carefully arranges can cause trouble because it can require more memory space than desired. The subject of when to use and when not to use "alignment" is highly compiler and hardware dependent. Therefore further specifics are left to the reader and his/her local reference material for clarification. Alignment is included here only for completeness.

EXTERNAL DATA BASE

Although the internal data base can affect a program's design, it primarily affects the storage requirements and cpu cycles. Selection of an external data base organization *can* radically affect program logic design. For example, the design logic necessary to update a master inventory file using sequential file logic is quite different from the logic needed to update the same file using various random access techniques.

To begin, let us define what is meant by an "external data base." Essentially, it is the logical *external* extension of the internal data base (i.e., data elements or variables). Usually these data elements are grouped into logical or physical structures (or records) which are maintained as data collections or sets often called files. These external and associated external data elements can have the same storage attributes as internal data elements, and external data elements occupy storage in the same manner as their internal counterparts. It is the retrieval and storage of these external structures or records from files into main memory that affects the program design. To help understand this bond between the external, physical data organization and the internal program logic design, let us review a few typical external data access techniques. From this introduction, the reader should have a rudimentary "menu" of choices with which to match various applications.

Sequential Access

Sequential access is well known to most programmers. Cards, tapes, and printed output are all forms of sequential processing. Using

Figure 8.5: Typical Sequential Processing Algorithm

SEQ_ACCESS: PROC;
   GET 'first record'
   PROCESS_LOOP: DO WHILE 'more_records';
     'process the record'
     GET 'next record'
   END PROCESS_LOOP;
END SEQ_ACCESS;

---

sequential access, a typical programming algorithm is to read the first record and (if present) begin a loop to process that record, then attempt to read the "next" record, and so on until all of the records have been read and processed. This algorithm is illustrated in Figure 8.5, using a sample set of pseudo-code.

Using sequential modes of input/output, no complex program design constraints are required and it is this simplicity which makes this approach so widely used. However, as the volume of records to be processed increases so does the time required to process data sequentially. Eventually, sequential access techniques cannot provide adequate throughput, and other techniques are needed.

Direct Access

Direct access differs from sequential access in its ability to "directly" access a unique record desired from a file without the sequential overhead of examining all prior records. Direct access files are arranged (or ordered) on a specific unique "key" which can be a single data element or a concatenation (stringing together) of several data elements. The key is used to access, insert, and remove records from the file. The typical processing algorithm used in an application program is to first obtain a key and then either access or store the desired record based on that key.

A direct access file can be conceptually viewed as an external array of structures with the keys functioning as subscripts. Key selection techniques have a definite impact on the storage requirements of a direct access file. For example, an individual's social security number may be an excellent "key" to reference an employee file, although allocating a unique "key" location for every possible key value (000-00-0000 through 999-99-9999) on the direct file would require file space for one billion records! Fortunately, there are other ways to

Figure 8.6. Employee Training Record

```
DECLARE 1 TRAINING_RECORD
    2 EMPLOYEE_NAME        CHAR(30),
    2 EMPLOYEE_NUMBER      PIC '(5)9',
    2 TRAINING_HISTORY     (10),
        3 CLASS_TITLE      CHAR(30),
        3 CLASS_DATE       PIC '(6)9',
        3 INSTRUCTOR       CHAR(20);
```

access a direct file within a manageable subset of the records required for the unique "key" configuration.

Typically, this is accomplished in one of two ways. The first is to provide a "hashing" algorithm for the key. This simply converts (or maps) the key into an address within the desired subset of records reserved on the direct file. Whenever the "hashing" algorithm generates the same address for two different keys, "a collision" occurs and an additional algorithm is required to produce a second (and possibly third) address until the desired record is either located or stored. Usually hashing and the collision handling algorithms require extensive programming logic plus extra processing overhead.

The second method of mapping keys into a subset of their potential range is to group the logical records (in sequence) into larger physical records (or blocks), and maintain a "directory" of the highest key value contained in a block. Access is achieved by first examining the "directory" for the block that *could* contain the desired key. That block is then retrieved (via a simple directory key) and the records contained in the block quickly searched for the desired key. Storage is accomplished in the same manner, except that space must be reserved in the blocks for additions, and when reserved space is occupied additional storage locations must be acquired and recorded in the "directory." This type of access (unlike hashing) is usually provided as *standard* software with most computer vendors, e.g., IBM's ISAM and VSAM.

Linked Lists

In designing the external data base, there are occasions when a single fixed record type is not satisfactory for an application program. For example, a single record used to maintain employee training records is illustrated in Figure 8.6. The problem with this approach is that the

A Primer on Structured Program Design

Figure 8.7: Simple linked list

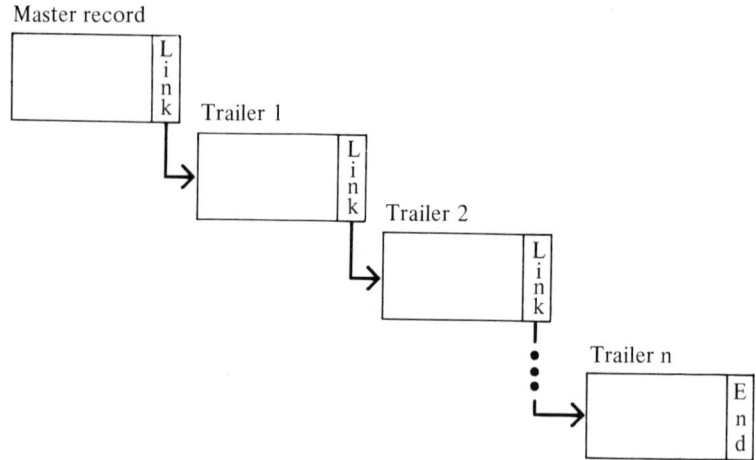

space allocation is likely either too big or too small for any given employee. If the record is too big, storage space is wasted. If the record is too small, some of the information desired will be lost.

The linked list accessing scheme can be used to solve this problem. Figure 8.7 illustrates the logic of the method. Here the basic idea is to maintain a master record (sometimes called a header or master link) containing the pointer or "link" to the first detail link (sometimes called a trailer). Each trailer contains not only additional data, but also a link to the "next" trailer record. Each record points to its next related record until the final trailer contains a special link indicating the "end-of-chain." Using the linked list logic, space allocation problems are solved with only a slight storage overhead of the "links" in each record.

Using the concept of a linked list, it is possible to create quite complex data structures. Figure 8.8 illustrates an employee personnel data structure which has several different "lists" of trailer records linked to a single master record. This type of data structure is called a multiple linked list.

Linked list data structures are generally used in conjunction with direct access techniques with the linkage number being an access key value. To assist in the maintenance and recovery of a linked list, additional pointers are sometimes used. Generally, these pointers

Data Base Considerations

Figure 8.8: Multiple linked lists

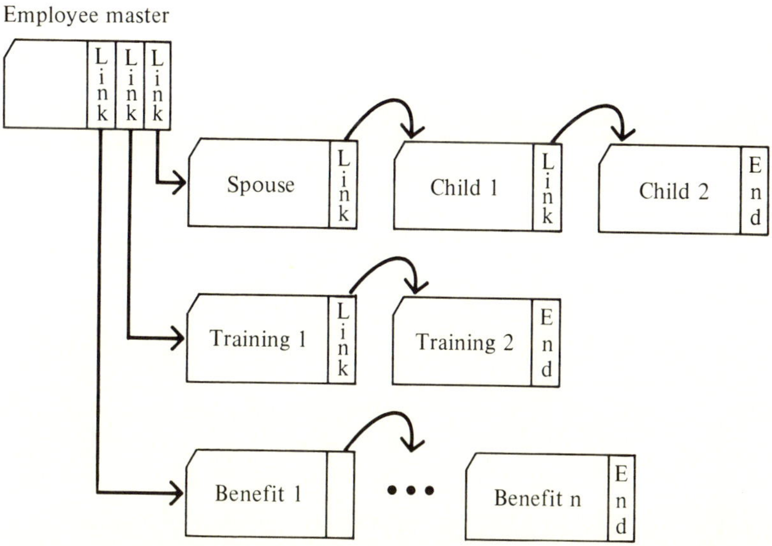

Employee master

[Diagram showing an Employee master record with three Links branching to three linked lists:
- Spouse → Child 1 → Child 2 (End)
- Training 1 → Training 2 (End)
- Benefit 1 → ... → Benefit n (End)]

provide a path from a trailer link back to either the master record, the previous trailer record, or both. In the event that some intermediate link is destroyed, the file can be restored minus only the "missing link." (Note: contrast this to the effects on file restoration without such a link.) By pointing back to the previous trailer record, the task of addition and deletion of links is made easier at the expense of additional storage for the pointers in each record. Thus the problem of data base recovery adds additional complexity and processing overhead to the design concepts illustrated here. Figure 8.9 illustrates a multi-linked file.

For the programmer, linked lists require an additional level of complexity in the application program logic to access and maintain the links. In addition, if both the master records and trailers are maintained in a single direct file, the programmer must also assume the responsibility for mapping the keys and pointers into unique addresses. Normally, a simpler approach is used. By separating the master records and trailers into two separate files, a simple direct access approach can be used for the keyed master records and a simple direct key (usually numeric) used for the trailer records.

Figure 8.9: Multilinked file

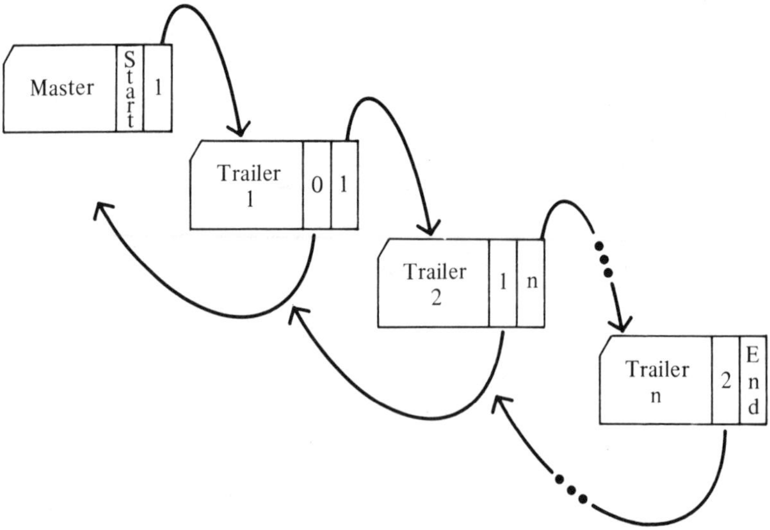

Inverted Files

Direct files are normally organized on a single key. If the desired data item is a function of that key, then retrieval is simple. However, if the desired data item is *not* a function of the key, then the entire file must be searched to satisfy the retrieval request. This is also true of both keyed direct and linked list files. For example, a state's motor vehicle file may be quite efficiently organized by vehicle registration number for administrative purposes, but police requests are *not* of this type. More likely a police request will be by license number or "blue 1970 Ford Maverick." To best solve this type of inquiry an access technique known as "inverted files" is used. Inversion involves the creation and maintenance of a separate direct file "directory" which cross-references secondary items. These secondary lists are organized by specific data values for the desired access to the secondary key fields. For example, the "color" field in the vehicle registration file would have keyed entries for "red," "green," "blue," etc., while the "model year" field would have entries for each year. If every data field is represented in this directory, the file is said to be "fully inverted." Generally, only a few fields are selected for inversion due to the extra storage and processing overhead incurred. These files are said to be "partially inverted."

Data Base Considerations

Figure 8.10: Inverted directory organization

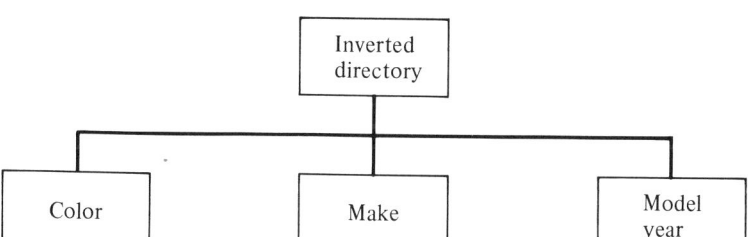

Figure 8.10 demonstrates how an inverted directory can be organized. In this illustration, each inverted field has a unique directory keyed on the field values. Each record in the directory contains a list of record keys in the master data base corresponding to records containing that field's directory value. To locate all "red Ford station wagons for 1973," the inverted directories for color, make, and model year are examined to find the "key lists" for "red," "Ford," "1973," and "station wagon." Once these lists of keys are located, the key values for the fields (starting with the shortest list) are compared to produce a list of keys in the master data base that satisfy the search criteria. This list will contain *only* the keys of records in the master data base that satisfy the search criteria.

The reason that the inverted file concept is popular lies in its speed for retrieval of selected items versus the time required to search the entire master file. With an inverted list, the bulk of the searching can be accomplished using only the key comparisons, thereby eliminating time-consuming record retrievals. As a result of handling only keys, just the records that meet the original search criteria need to be actually retrieved from the master data base. The programming costs for this quick access can be significant, but in some high retrieval environments, such as auto registration systems, logic such as this is necessary.

Directories for the inverted fields have to be established and maintained. Additions, changes, and deletions to the master data base must correspondingly add, change, and delete entries in the "inverted data base." As a result, inverted files are normally reserved for use with only the largest of data bases, which often are also prime candidates for commercial data base management systems.

SUMMARY

Program design reflects the selection of both an internal and external data base. Internal variables are restricted to the type of data that can be stored by associating specific variable types or attributes with each variable. These variable types specify the data type, aggregate type, and alignment type. Arithmetic storage can be defined many different ways (ignoring differences in scaling and precision). It can be efficiently manipulated by the cpu, but requires conversion for input and output. String storage is divided into two basic classes—ordinary and pictured. Ordinary string storage is used primarily for text and boolean variables, while pictured string storage is used for numeric values. Pictured string storage provides an alternative to both ordinary and arithmetic storage types and is primarily useful in applications which are characterized by high-volume input and output with few arithmetic calculations.

Scalar variables can be aggregated as either structures or arrays. Structures are nonhomogeneous sequences of numbers. Each member of a structure can have any storage type, i.e., arithmetic, string, pictured, etc. An array variable is a sequence of homogeneous elements sharing a single data name and storage type. Using subscripts, general expressions for an array allow specific elements of an array to be dynamically accessed during processing. An array can specify any storage type *except* another array. All internal variables are either aligned or unaligned in storage or memory. Compilers generally default to aligned storage to improve program execution efficiency; however, this *can* cause an excessive requirement for the actual storage needed by the program's data. By carefully specifying the unaligned attribute and by specific ordering of the data elements, a programmer can reduce the storage space required—at the cost of execution efficiency.

Normally, the internal data base selection affects the storage requirements for a program and the external data base selection can radically affect the program design logic. The external data base is a logical extension of the internal data base. It is organized into logical structures or records and these structures are contained in files or data sets. The external data elements have the same attributes as internal data base elements, i.e., arithmetic, string, etc. However, it is the accessing of the external data base that affects the program logic design. Sequential access is the simplest form, is well known to

Data Base Considerations

programmers, and requires no complex logic. Direct access differs from sequential access in its ability to process *only* the desired record from a file without the sequential overhead of examining all prior records. Direct access files make use of a unique "key" field for storage and retrieval. Programming techniques to use direct access are simple, but can result in unnecessarily large file storage allocations to accommodate all possible key values (e.g., social security number). Fortunately, methods exist to reduce the actual storage space required, while the ISAM method is usually provided by the cpu vendor.

Linked lists can be used to store and access records with varying amounts of additional data. Each record is "linked" to the next associated record, from the master record to the last record in the "chain." Linked lists can be used to create quite complex data structures.

The use of inverted files represents a technique for providing direct access to records based on a secondary key value. When using an inverted file, secondary keys can be used to select the desired records to be retrieved. This directory comparison eliminates retrieval of unneeded records and speeds the overall retrieval process.

The choice of external file organization can impact the design logic many different ways (i.e., I/O recovery techniques, data manipulation, etc.). Careful consideration should be given to the level of complexity introduced by a specific access technique versus the benefits that will be provided. Also, potential growth of a data base can effect future access performance.

Review Questions

8.1. What are the three types of typical variable specifications?

8.2. Does the language that you normally use for programming allow all eight types of arithmetic variables? If it does not, then which type (or types) would you include if given the opportunity? Why? If your language does allow all eight choices, which type (or types) would you omit if given the chance?

8.3. Why shouldn't every string variable be "declared" with the VARYING option?

8.4. If you are designing a program with a very complex arithmetic formula and the program is to process a large number of inputs and produce an even greater number of outputs, what variable type (or types) would you select? Why?

8.5. What is the difference between a scalar variable and an aggregate variable?

8.6. What advantages do homogeneous aggregates have over heterogeneous aggregates? What disadvantages? When might you select one over the other?

8.7. If you were to design a direct access program to process an inventory accounting system which used supply numbers as the principal retrieval key, would hashing be the best method of implementation?

8.8. Using linked lists, a number of considerations should be considered in the design of the list structure, i.e., additional links. To illustrate the importance of considering these additional links in the initial data design, first briefly pseudo-code a simple linked list file and then modify the pseudo-code to handle multi-links.

8.9. What are inverted files? Why are they considered for an application?

Notes and References

1. Jackson, M.A. *Principles of Program Design.* New York: Academic Press, 1975.
2. Peters, L.J. and Tripp, L.L. "Comparing Software Design Methodologies." *Datamation* 23 (November 1977), pp. 89−94.
3. IBM Systems Reference Library *IBM OS Full American National Standard COBOL.* Palo Alto: IBM Corp., 1975.
4. Cress, P.; Dirksen, P.; and Graham, J.W. *FORTRAN IV with WATFOR and WATFIV.* Englewood Cliffs, N.J.: Prentice-Hall, Inc., 1970.
5. Kroenke, D. *Database Processing.* Chicago: Science Research Associates, Inc., 1977.

9
MANAGING THE PROGRAMMING PROCESS

Plan, organize, control—this is the essence of all management. Organizing is a relatively simple task that managers can easily comprehend and accomplish. But for the DP manager concerned with software development, effective *planning* and *control* appear to be elusive goals. The traditional approach to software development is typified by the "90% complete syndrome" described in chapter 1. As early as 1969, adequate control approaches had been identified as management failures in the software development process. Brooks perhaps states the problem of control best in his description of the prehistoric tar pit: "Everyone is surprised by the stickiness of the problem. . . ."[1]

No single task in software development seems insurmountable. Yet, taken together, all the many individual tasks and problems generate a synergistic effect, that is, the action of the separate tasks and problems generate an effect greater than the sum of the individual parts. Perhaps the difficulties arise because we, as programmers (and managers), don't fully understand what we are developing. Our planning criteria are too vague and our estimates of resource allocations are inadequate.

An individual program (or set of programs) is a component of some design structure or hierarchy. Whenever our attention is focused on a program as the *solution* to a problem, we have already lost sight of both the forest and the trees. By viewing the solution as a piece of software we are concentrating on one of the lowest elements of the hierarchy, i.e., the leaves. Viewing the solution to an application problem as a program is the "bottom-up" approach to software development that makes the task of management so difficult. To improve our chances of success (in software development), let us

Figure 9.1: Management control structure

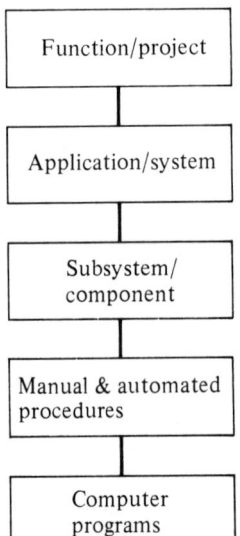

examine a management control structure which is outlined schematically in Figure 9.1.

This figure illustrates a hierarchy of components within an information processing structure. It is important to realize that the first four elements exist whether or *not* a computer is used. For example, accounting existed as an information processing *function* long before the advent of automatic data processing equipment. Within the accounting function, applications (or systems) for payroll, accounts receivable, and accounts payable existed long before the first computer program. And within these applications (or systems) people accomplished their individual tasks according to some form of established *procedures* long before the first punched card. Whenever automation or software is inserted into the hierarchy of Figure 9.1, it should be very apparent that it *supports, not replaces*, other components.

The "correct" approach to software development is as varied as the authors who publish articles and books (not to mention countless local, unpublished guides). One excellent software development methodology that is successfully marketed by M. Bryce & Associates, Inc., Cincinnati, Ohio is the PRIDE system. Other methods and techniques are available, but the essence of the PRIDE methodology is the enforced, early (and complete) end-user involvement in system development. Some experts estimate that one-third of all system changes occur because of poor communication between designer and user. PRIDE attempts to rectify this problem, plus define (plan) the system development in such a way that future control is simplified.

User involvement in the planning and development of a software project is *vital*. Except for the very small or unique organization, it is the end-user *not* the programmer that must use the software. Users generate the input, schedule the execution of the program(s), and examine the output—day in and day out. In most cases the users did these same tasks manually long before programs were created. Therefore, before software is considered as a solution to a specific problem, the user's function should first be examined to determine the specific application (or applications) that are affected. Then, the individual, time-phased components of that application need to be coordinated (with the user) to determine what and how the proposed software should best be developed and implemented. The PRIDE methodology provides a sequence of phases and steps to be accomplished as shown in Figure 9.2.

It is beyond the scope of this chapter to fully describe the PRIDE development techniques for a system of programs. However, we can describe how to manage the development of a *single* program. It is hoped this management process will illustrate the practical use of the various structured programming techniques described in detail in the previous chapters.

BASIC COMPONENTS OF PROGRAMMING MANAGEMENT

Managing the programming process incorporates several of the more familiar components of structured programming. These components are the structured walk-through, HIPO charts, pseudo-code, top-down functional design, top-down development, and, of course, structured code. The concepts of the formal chief programmer team and the development support library are deliberately excluded since both assume a specific organizational structure which may or may not be easily implemented at a given data processing center. If program management consisted of nothing more than implementing the above mentioned "structured programming" elements, it would assist

Figure 9.2. System Development Phases and Related Documentation

	Phase	Related Documentation
1.	System Study and Evaluation	System Study and Evaluation Report
2.	System Design	System Design Manual
3.	Sub-System Design	Sub-System Design Manual
4a.	Administrative Procedure Design	Administrative Procedure Manual
4b.	Computer Procedure Design	Computer Run Book
5.	Program Design	Program Documentation
6.	Computer Procedure Test	Additions to the Computer Run Book
7.	System Test	Additions to the Systems Design Manual
8.	System Operation	Modification Improvement Requests
9.	System Audit	System Evaluation Report

Note: Condensed from PRIDE® Methodology with the permission of M. Bryce & Associates, Inc., Cincinnati, Ohio.

Managing the Programming Process

Figure 9.3: Program development checklist

TOP-DOWN PROGRAMMING
WORKSHEET

PROGRAM NAME: _____
PROJECT CODE: _____

MOD NO.	Modules Accepted Initials & Date	Development Stage and Test Plan (Includes Test Data Plan)								Remarks
	Source Code Test Stub Full Cases	1	2	3	4	5	6	7	8	
	Development Stage Test Results Accepted (Initials & Date by Designated Reviewer)	1	2	3	4	5	6	7	8	

individual programmers in the production of structured, error-free code, and improve some internal project planning and organization. But this chapter's objective is to describe a simple method to manage the development of a single program. The technique chosen to accomplish this is the Program Development Checklist.

Program Development Checklist

The Program Development Checklist (Figure 9.3) is used to define

A Primer on Structured Program Design

Figure 9.4: Checklist step 1

Program Development Checklist

PROGRAM NAME: **Input Edit**   TARGET DATE: **31 Oct**
PROJECT CODE: **KD 22C**
CHIEF PROGRAMMER: **J. Thomlinson**
REVIEWER: **B. Dietz**

	Programming milestones / Work Products	Objectives	Milestone Acceptance (initials and date)		
			Reviewer	Manager	Other
1	Program specifications	Clarity, completeness	/////	/////	24 Aug. Chief Ap. Programmer
2	Functional Program design and projected completion date **15 Oct.**	Structured pgm design, easily modifiable	Dietz		

and track a program through the development cycle. The checklist could be used to manage many different-sized programming tasks, from a single subroutine to a complete operating system. However, its principal focus is the individual program, since this is a universal, easily understood unit of work. The checklist chronicles the work that has been accomplished and shows what program development phase is scheduled to be accomplished next. The checklist is a *management* tool to control and record the program development process. It is a step-by-step schedule of how a program is to be developed and a record for management showing the acceptance of each development step. Implicit in this process is authorization for the programmer to proceed to the next step. In addition, the checklist serves as a record of personnel assignments for a program. Using the Program Development Checklist, a programming manager has a vital tool to manage the "structured" development of a program.

PROGRAM DEVELOPMENT CYCLE

Once a manager has a program to be developed, the initial step (Figure 9.4) is to assign the program development to someone (even himself/herself). This is accomplished on the checklist by completing the control information at the top of the form and designating the assigned person as chief programmer. This first step includes the vital

Managing the Programming Process

Figure 9.5: Checklist step 2

Functional program design and projected completion date	Structured pgm design, easily modifiable	As 11 Sept.	To Winkler 12 Sept.	

face-to-face contact with the assigned programmer to clarify the specifics of the assignment and to obtain a commitment (and even more important, acceptance) to the delegated task.

Therefore, the first step of the checklist is complete *only* when the designated chief programmer is satisfied that *he/she* understands the task to be performed. Obviously, the successful completion of this step does not mean to preclude or prevent additional or clarifying specifications at a later time, nor does it imply that this step can always be accomplished in a single meeting. The sole objective of the first checklist step is a thorough understanding of the task at hand by the person assigned that task.

The second checklist step (Figure 9.5) is the development of a preliminary top-down functional design of the program and a projected completion date. The key element of this step is to obtain a basic, structured design that can be implemented on time, without logic errors, and be easily enhanced at some future date. Top-down functional design has been the subject of many previous discussions (primarily chapter 5). It is important to remember at this point in the program development process that the top-down design concept focuses the programmer's attention on the design *structure*, and has as an objective simplifying that structure. Brooks comments on this aspect in the following excerpt:

> The important point, and the one vital to constructing bug-free programs, is that one wants to think about the control structures, of a system, as control structures. . . . This way of thinking is a major step forward.[1]

Based on the top-down functional design, the programmer must then estimate the number of lines of code required to implement the design. Initially, this can be some arbitrary figure such as the number of lines on a single page of source listing (i.e., 25–50 lines multiplied by the number of separate functions or modules identified by the top-down design process). With experience this arbitrary value can be refined.

A Primer on Structured Program Design

Figure 9.6: Assignment of reviewers

Program Development Checklist

PROGRAM NAME: __Input Edit__   TARGET DATE: __31 Oct__
PROJECT CODE: __KD 22C__
CHIEF PROGRAMMER: __J. Thomlinson__
REVIEWER: __B. Dietz__

	Programming milestones Work Products	Objectives	Milestone Acceptance (initials and date)		
			Reviewer	Manager	Other
1	Program specifications	Clarity, completeness	//////	//////	24 Aug. Chief Ap. Programmer
2	Functional Program design and projected completion date __15 Oct__.	Structured pgm design, easily modifiable	Dietz		
3	HIPO package and top-down development plan	Specifications met readability, and functional dsgn confirmed		G. Andrea	
4	Top-down development (use worksheets on reverse side)	Top-down coding and testing of the program	Dietz	G. Andrea	
5	Completed program	Thoroughly tested program	Dietz	G. Andrea	B. Zelonis
6	Program documentation	All applicable program documentation rqmts completed	Dietz		
	PROJECTED COMPLETION DATE: __15 Oct.__				
	PROJECTED IMPLEMENTATION DATE: __1 Nov.__				

Additional Personnel
Assignments: DEVELOPERS   REVIEWERS

Primary Backup: __Healy__
Team Members: __Elliot__

(Note: For estimating purposes, 30 lines per module provides a reasonable estimating value.) Once the total number of source code lines has been estimated, the projected completion date is estimated by dividing the source code lines by a historical or projected programmer productivity standard (e.g., 30 lines of source code per

Managing the Programming Process

programmer day). Again, the accuracy of this estimating technique will improve as both programmers and managers become accustomed to the structured programming techniques explained in this book. In fact, the collection of historical data on various development projects will aid in producing accurate estimates.

The primary function of the projected completion date is to alert management early in the development process of *potential* delays. If there is a difference between this date and the manager's target date for the project, the manager has *early warning* of this difference and the option of either changing the target date or expanding the resources available to the effort.

Should the manager decide to use a programming team to develop the program (and a team can be two or more), their selection can be recorded on the checklist in the appropriate section. At this phase of development, the manager must consider the review process for the subsequent program development steps. Some managers maintain very close control of the project at each specific phase or step, while others may elect to delegate complete review responsibility to the team. In either case the checklist provides a place to list the reviewers and their roles at the various program development checkpoints (Figure 9.6).

The reviews of each step are accomplished via a structured walk-through. These formalized sessions concentrate on problem detection. During a structured walk-through, the chief programmer verbally presents his/her work product for that phase to a group of peers and reviewers. The basic purpose of a structured walk-through is to detect design or coding errors at the earliest point in the development cycle when the cost of correction is minimal. A checklist step is considered complete *only* when the reviewers are satisfied that the approach is adequate and technically accurate. It should be emphasized that each checklist step is a development milestone in a program's development. Their primary purpose is to focus the team's attention to each phase, in turn, and to proceed to the next step *only* when the team has satisfied the designated reviewers that they have accomplished the current milestone.

The third step in the program development checklist is the formal transformation of the functional design into HIPO charts and pseudocode which can be used as design documentation. Initially, for programmers, this step is analogous to the castor oil impact of flowcharting or some other dread managerial paperwork curse.

Phillip Metzger describes this feeling as follows:

> People gag at the idea of paperwork because they are impatient to get on with the job. . . . Resist the temptation to begin (coding) immediately. Concentrate first on WHAT the problem is, NOT on HOW you are going to solve it.[2]

HIPO charts are a relatively new technique for describing input- and output-oriented systems, and are used in the program development checklist to replace flowcharts as the primary system and logic documentation vehicle. Many contemporary system designers believe that the flowchart is oversold as a technique of program documentation. In structured design, programs use flowcharts only to illustrate broad system flows. Flowcharts tend to emphasize the decision structure of a program, which is only one aspect of its design. They cloud decision structure rather elegantly even when the flowchart is on one page, but even this overview breaks down badly when the flowchart has multiple pages sewn together with numbered exits and connectors. In addition, who uses flowcharts to maintain any functional production program?

HIPO charts are included in the production development checklist because they provide a "natural" transition from functional design to actual coding. HIPOs and pseudo-code are both critical components of structured programming techniques. Once the functional design has been transformed into HIPO charts, the entire program can be tested on paper prior to coding for basic control logic errors, missing or improperly defined functions—all while the impact of such errors on the project is minimal. During the maintenance cycle, these original HIPO and high-level pseudo-code documents will provide a valuable source of documentation to quickly identify and implement changes and modifications. Again, checklist step three is not finished until a structured walk-through convinces reviewers that the development package meets program specifications, plus a readable and comment package meets program specifications, plus is readable and complete. At the conclusion of the third step, the program design is finally ready to be committed to code.

Step four, top-down program development, uses the reverse side of the program development checklist to control the program's detailed transformation into code. Implicit in the acceptance of step three is a skeleton worksheet (Figure 9.7) outlining the plan for top-down programming. The worksheet now becomes a coding and testing schedule. The program is arbitrarily broken into phases and each is

Figure 9.7: Program development worksheet

TOP-DOWN PROGRAMMING WORKSHEET

PROGRAM NAME: **Input Edit**
PROJECT CODE: **KD 22C**

MOD NO.	Modules Accepted Initials & Date / Source Code Test Stub Full Cases	1	2	3	4	5	6	7	8	Remarks
10		Full								2: Healey
20		Stub	Full							2: Elliot
30			Stub	Full						2: Healey
40			Stub	Full						2: Tomlinson
50			Stub	Full						2: Tomlinson
60		Stub	Full							2: Tomlinson
70			Stub	Full						2: Healey
80			Stub	Full						2: Elliot
90				Stub	Full					2: Healey
100		Stub	Full							2: Tomlinson
110			Stub	Full						2: Elliot
120			Trace	Stub	Full					2: Tomlinson
130				Stub	Full					2: Tomlinson
140				Stub	Full					2: Healey
900		Full								2: Tomlinson
	Development Stage Test Results Accepted (Initials & Date by Designated Reviewer)	1: 8/29 BD	2: 9/15 BD	3: 9/30 BD	4: 10/15 BD	5	6	7	8	

coded and tested in a top-down fashion. The worksheet provides a control point for the chief programmer (and team members) for module coding assignment, particular modules required for a specific test phase, and the level of coding required for a given module at any given test phase. The level of code describes either a fully coded module (complete), a stub module, or trace module. Stub modules generally return pseudo-conditions or dummy variables when called. Trace modules simply print their names on an audit listing to provide a debugging flow of control audit trail. Both stub and trace code initially provide nonfunctional subroutines which can be called by other more complete functions. In addition, trace code, depending on the compiler language options, can remain in the completed program, while stub code is always temporary, throw-away code.

The program development worksheet is not simply a convenient place for project bookkeeping; it is the primary vehicle for planning and implementing the top-down development of the program. Some of the major benefits of top-down development are summed up by Edward Yourdon in the following list:

a. System testing, in its classical sense, is virtually eliminated, since each execution essentially is a system level test.
b. Major interfaces are tested first. As a result, major bugs are discovered early . . . while trivial bugs are discovered toward the end.
c. Users can be given a preliminary version of the program at a relatively early stage.
d. If it is not possible to finish the entire program by the time a deadline has arrived, it is likely that a useable sub-set of the program will be finished.
e. It is often much easier to find bugs with a top-down testing approach.
f. Testing time is distributed more evenly throughout the project, thus eliminating the requirements for large amounts of computer time toward the end of the project.
g. The programmers' morale is improved considerably when they can see the results of a successful test of the final program skeleton.
h. Top-down testing provides a natural vehicle for testing lower level modules.[3]

The key to a successful top-down development phase is in the planning. Probably the best way to plan top-down development is to postpone the creation of those modules which require the most test data until late in the top-down plan. Thus, each new phase can use the test data from the previous phases with only the new test data required for the current phase being tested. This method of program development has the added advantage of permitting the basic program control logic to be thoroughly debugged without the unnecessary overhead of concurrent debugging of each detail module. Therefore, by the final phase, a complete file set of both good and bad test data has been generated. Although the program development worksheet provides for eight separate test phases, it will be a rare program that will require more than three or four separate phases.

The concept behind top-down development is to develop and test the program in phases and not as a single monolith. As each phase is completed, the test results are accepted by the reviewers, and the programming team then proceeds to the next phase, and the next, until step four of the program development checklist is complete.

Step five is the final structured walk-through of the completed program, with all its products and a complete set of the test data. Approval of this step is formal verification that the program is accepted by the users and ready for production implementation.

Step six is the final program documentation effort. One major advantage that the program development checklist offers is to force documentation to become a by-product of the program development effort. The primary goal in step six is to have the remaining documentation effort which involves assembling the various component parts into an acceptable program documentation folder/package, plus updating the top-down design charts to reflect any minor changes that occurred during the debugging and testing phases. It is suggested that documentation requirements be restricted to only the absolute essentials, relying primarily on machine-produced products such as compiler listings with cross-references, test runs, HIPO charts, and pseudo-code. The objective in step six is for the chief programmer to have *very little* to do to finish the assigned program. Again, as always this final step is not complete until accepted by the reviewers via a structured walk-through.

SUMMARY

Management of a program is basically *not* a complex task. It involves the simple application of a logical, systematic approach to the task. The Program Development Checklist provides a convenient tool to improve management visibility and control. The principal value of the checklist is to bring *active* involvement of management into the development process. Through structured walk-throughs each phase of a program's development is reviewed and compared to the specifications. It is through these reviews that logic errors are identified for correction. Perhaps more important than logic errors is the detection of logical omissions and extraneous functions. When used diligently the checklist enforces an orderly approach to program development; however, implementation of the checklist requires affirmative action by management to *do* something about program development. The checklist requires a manager to perform management functions which are often alien to the data processing field. We believe the software development process is essentially a similar activity to many other project-oriented operations, not a mysterious act to be accomplished behind locked doors. The cost and organizational severity of this effort *demands* that it become *more* visible.

Problems

9.1. In the next program that you develop use the checklist as shown in this chapter to assist in the various programming development steps. Do not proceed from one step to the next until *someone else* has reviewed your work and understands what you have produced.

9.2. The next time you have a program that is too big for a single person to comfortably develop alone, use the checklist to help the team of programmers develop the program. What benefits did you observe? How can you improve the checklist to fit your own particular needs?

9.3. To demonstrate the potential of the structured programming concepts described in this book, use these techniques (with a team of programmers) to develop any application in your organization that is: (a) recognized as needed, and (b) continually postponed due to lack of personnel expertise or time.

Notes and References

1. Brooks, Frederick. *Mythical Man-Month: Essays on Software Engineering.* Reading, MA: Addison-Wesley Publishing Co., 1975.
2. Metzger, Phillip W. *Managing a Programming Project.* Englewood Cliffs, N.J.: Prentice-Hall, Inc., 1975.
3. Yourdon, Edward. *Techniques of Program Structure and Design.* Englewood Cliffs, N.J.: Prentice-Hall, Inc., 1975.

10
FINAL REMARKS

We have covered a rather broad brush of our subject in this text and as with all such discussions much has been glossed over. The serious reader should review the works of Baker, Brooks, Constantine, Katzan, Mills, Myers, Orr, Yourdan, and others to develop a degree of depth in the various subject areas which were discussed here.*

Every reasonable attempt has been made to describe practical approaches to the subject by miniature examples and checklists. It is hoped that the checklists presented can be modified to fit specific local needs. We also hope that the material discussed is comprehensible as a tutorial for students or new employees desiring to learn more about the subject of structured programming and design.

Where is this subject going to evolve in the future? That is a difficult question to answer. Undoubtedly the traditional high-level languages will remain in broad use for some time, and data base integration will continue as computer mass storage costs decrease. Also, higher level retrieval languages are starting to appear in production use. Some feel that programs will be so easy to write that few of them will be saved (e.g., throw-away code). As computer users are able to construct programs using higher levels of logic, the wisdom of the currently conceived structured program blocks may be replaced by other logic constructs at a higher level of detail. At some future time one might be able to simply walk up to a retrieval device, give it a friendly pat (for identification), ask it for information verbally, and the answer will appear (verbally or on a video screen). Maybe this will occur, but not for some time, we suspect.

The microcomputer revolution has already started and, unfortunately, old problem patterns are being repeated. Many users do not

*See end-of-chapter references for specific details concerning these and other authors.

remember the problems which caused the structured programming revolution to being initially—poorly written software and the relative cost of human manpower required to accomplish the task. Today many software packages are now being written in BASIC (Basic All-purpose Symbolic Instruction Code) which was initially designed to motivate novices to use a computer. It does not facilitate structured design as described here and thus represents a future problem. If "good" structure is to be the goal of future systems, then new languages are needed which support this goal. If the structured programming revolution is to continue at its present momentum, we should begin to hear more of the new languages ADA, PASCAL, PL/M, and PL/Z, to mention but a few. Also, COBOL and FORTRAN will be modified for improved structuring capabilities. Whatever the language approach taken, the next frontier is the microcomputer and its attendant implementation problems in large and small organizations.

The driving forces for change in the future will be software development costs and growing expectations from the user environment. Management is going to increasingly expect visibility in a software design project and careful performance monitoring throughout. Only a structured approach to design can provide this capability. Early successes with structured programming and design techniques are reminiscent of those found with PERT (Program Evaluation and Review Technique) during the late fifties. PERT enjoyed great initial success, then disillusionment followed by regrouping of the implementation approach. Structured design will likely follow this sort of pendulum approach to acceptance.

The good news is that we now know structured design will work under certain selective situations. The bad news is that we don't know exactly what combination of tools and techniques will be optimal in a particular environment or situation. Only further research and time will answer these questions. As an example of this, HIPO generally works well in commercially oriented applications, but how about bit-oriented systems programming which often uses assembler and other programming tricks to improve execution speeds? Probably not so well here.

Can the chief programmer team concept be implemented, unaltered, in a civil service or other large bureaucratic organization? Probably not without considerable modification. There are, undoubtedly, motivational problems related to this entire process which are not well known or understood at this point. It seems reasonable to suspect that

Final Remarks

structured walk-throughs and general standardization of a previously exotic, free-wheeling job will cause further behavioral reactions. There are certain negative aspects to these and other related ideas of structured design. We may find short-term efficiency and long-term organizational resistance if management requires such measures to be implemented. It has been shown time and time again that no tool (management or analytical) laid over an otherwise poorly run operation will cause this organization to become effective. All of the concepts relate to organizational change which seldom comes without pain.

The set of structured design tools described here is essentially the third generation of design technology (see comments in chapter 3). We must remember that more generations are almost sure to occur and that the user spectrum spans a very wide range of applications and computer structures. No single aggregation of techniques will universally fit such a broad spectrum. Human intelligence must guide us through the maze.

APPENDIX: DEBUGGING CASE STUDY SOLUTION

The final biorhythm program listing and output of the debugging case study are presented in figures A.1 and A.2. The major program logic errors as well as minor style errors have been corrected.

All the variable names have been explicitly declared. Some of these names are marginally adequate (i.e., M1, M2, D1, D2, Y1, and Y2), while the remaining should be revised. The constants used throughout the program have also been assigned names, declared, and then initialized with these values. This is done primarily to reduce arithmetic conversion errors caused by mixed-mode operations.

A few original style problems remain. These should be the topics of further discussion.

DISCUSSION QUESTIONS

1. What could be the purpose of statements 58 and 62? Why would the author wish to know the number of iterations in the loop?
2. In statement 59, R receives the value of OH. R is again found in statement 61. Why was this change made from the original program?
3. Statements 15 and 16 could be combined to form one statement: P1=N7. Why would the author choose to use these two statements?
4. Statements 95 and 96 could be revised. The key to the revision is the value of F(2). How could these statements be more efficiently revised?

Other style problems may be present in this final program. The reader is left to analyze further the overall style.

Figure A.1: Biorhythm program listing—Version III

```
/*H  BIORHYTHM PLOTTING PROGRAM VERSION III */

STMT   LEV NEST        SOURCE PROGRAM TEXT

              #  /*H  BIORHYTHM PLOTTING PROGRAM VERSION III */
              #  BIO: PROC OPTIONS (MAIN);
              #  /*       BIORHYTHM FLOTTING PROGRAM        */
  1    1      #
  2    1      #     DCL Y                                    FLOAT(8),
              #         F(12)                                FLOAT(8)
              #         INIT(31,28,31,30,31,30,31,31,30,31,30,31);
  3    1      #     DECLARE
              #         T$(8)                                CHAR(3)
              #         INIT('SUN','MON','TUE','WED','THU','FRI','SAT',' ');
  4    1      #     DCL
              #         A(51)                                CHAR(1),
              #         K                                    FLOAT (8) INIT(6.283185),
              #         Z$                                   CHAR(20),
              #         (M1,D1,Y1,M2,D2,Y2,X,P1,P2,J1,J2)    FLOAT(8),
              #         (N1,N2,N3,N4,N5,N6,N7,Q,L,C)         FLOAT(8);
              #  /*PROGRAM CONSTANTS*/
  5    1      #     DECLARE
              #         V1900                                FLOAT(8) INIT (1900),
              #         V365                                 FLOAT(8) INIT(365),
              #         V4                                   FLOAT(8) INIT(4),
              #         V1                                   FLOAT(8) INIT(1),
              #         V23                                  FLOAT(8) INIT(23),
              #         V25                                  FLOAT(8) INIT(25),
              #         V26                                  FLOAT(8) INIT(26),
              #         V33                                  FLOAT (8) INIT(33),
              #         V28                                  FLOAT(8) INIT(28),
```

```
*                              V3                    FLOAT(8) INIT(3),
*                              V6                    FLOAT(8) INIT(0.6),
*                              V13                   FLOAT(8) INIT(13),
*                              V12                   FLOAT(8) INIT(12),
*                              V5                    FLOAT(8) INIT(5),
*                              V100                  FLOAT(8) INIT(100),
*                              V400                  FLOAT(8) INIT(400),
*                              (TEST1,TEST2,TEST3)   FIXED,
*                              END_FLAG              BIT(1) INIT('0'B);
 6       1       ON ENDFILE(SYSIN) END_FLAG = '1'B;
         1       /*   DATA INPUT SECTION    */
 7       1       GET FILE(SYSIN) LIST(ZS);
 8       1   LOOP3:
 9       1   *   DO WHILE (END_FLAG= '0'B);
10       1   *   PUT FILE(SYSPRINT)PAGE EDIT('ENTER DATES IN THE FORMAT MM.DD.YYYY'
                 )
                 (COL(25),A)
                 ('EXAMPLE-- JUNE 16.1944 WOULD BE 06.16.1944')
                 (COL(25),A);
11       1   *   GET FILE(SYSIN) LIST (M1,D1,Y1);
12       1   *   IF Y1<=99 THEN Y1=Y1+V1900;
13       1   *   M2=M1;
14       1   *   D2=D1;
15       1   *   Y2=Y1;
16       1   *   CALL DAY;
17       1   *   P2=N7;
18       1   *   P1=P2;
19       1   *   PUT FILE(SYSPRINT) EDIT ('BIRTHDATE:',M1,'/',D1,'/',Y1,TS(P1))
20       1       (COL(25),A,COL(35),F(2,0),COL(37),A,COL(38),F(2,0),COL(40)
                 ,A,
21       1       COL(41),F(4,0),COL(48),A);
22       1   *   GET FILE(SYSIN) LIST (M2,D2,Y2);
```

/*H BIORHYTHM PLOTTING PROGRAM VERSION III */

```
STMT  LEV NEST      SOURCE PROGRAM TEXT

 23    1   1    *     IF Y2<=99 THEN Y2=Y2+V1900;
 24    1   1    *     CALL DAY;
 25    1   1    *     P2=N7;
 26    1   1    *     PUT FILE(SYSPRINT) EDIT ('START DATE FOR CHART:', M2,'/',D2,'/',
                                Y2,
                                T$(P2))
                               (COL(25),A,COL (50),F(2,0),COL(52),A,COL(53),F(2,0),COL(
                                55),A,COL(56),
                                F(4,0),COL(65),A);
 27    1   1    *     GET FILE(SYSIN) LIST (L);
 28    1   1    *     PUT FILE(SYSPRINT) EDIT ('LENGTH OF CHART IN DAYS:',L)
 29    1   1    *                       (COL(25),A,COL(55),F(2,0));
 30    1   1    *  /*    CALCULATE NO OF DAYS TO START DATE, TAKING LEAP YEARS INTO  */
                /*                                                  ACCOUNT          */
 31    1   1    *     X=M1;
 32    1   1    *     CALL JULIAN;
 33    1   1    *     J1=J2+Y1*V365+D1;
 34    1   1    *     IF J1<639723 THEN P1=8;
 35    1   1    *     X=M2;
 36    1   1    *     CALL JULIAN;
 37    1   1    *     J2=J2+Y2*V365+D2;
 38    1   1    *     IF J2<639723 THEN P2=8;
 39    1   1    *     OH=J2-J1+FLOOR((Y2-Y1)/4+.499999);
 40    1   1    *     PUT SKIP LIST ('*** OH VALUE ***');
 41    1   1    *     PUT SKIP DATA (OH,J2,J1);
 42    1   1    *     IF Y1/V4=FLOOR(Y1/V4) =0 THEN DO;
 44    1   2    *        IF Y2>Y1 THEN
 45    1   2    *           IF M1<V3 THEN OH=OH+V1;
```

```
46  1    #            IF M2>2 THEN
47  1    #              IF M1<3 THEN OH=OH+V1;
48  1    #                                                END;
49  1    #    PUT FILE(SYSPRINT) EDIT ('OH= ',OH)
50  1    #                           (COL(120),A,COL(126),F(7,1));
         /* PRINT CHART HEADER */
51  1    #    PUT FILE(SYSPRINT) PAGE EDIT ('BIORYTHM CHART FOR',Z$)
52  1    #                           (COL(50),A,COL(75),A);
53  1    #    PUT FILE(SYSPRINT) EDIT('BORN ON  ',T$(P1),M1,'/',D1,'/',Y1,
                                'BEGINNING',T$(P2),M2,'/',D2,'/',Y2)
                    (COL(54),A,COL(64),A,COL(68),F(2,0),COL(70),A,COL(71),F(2,
                    0),COL(73),A,
                    COL(74),F(4,0),COL(54),A,COL(64),A,COL(68),F(2,0),COL(70),
                    A,COL(71),
                    F(2,0),COL(73),A,COL(74),F(4,0));
54  1    #
55  1    #    PUT FILE(SYSPRINT)SKIP(2) EDIT ('P=PHYSICAL            (23 DAYS)',
                                'E=EMOTIONAL           (28 DAYS)',
                                'I=INTELLECTUAL        (33 DAYS)',
                                'A=OVERALL AVERAGE')
56  1    #                (COL(54),A,COL(54),A,COL(54),A,COL(54),A);
57  1    #    PUT FILE(SYSPRINT) SKIP(3) EDIT ('DOWN','CRITICAL','UP',
                    '                                                       ')
58  1    #                (COL(42),A,COL(63),A,COL(92),A,COL(42),A);
         /* SET F(2) TO 29 FOR LEAP YEARS*/
59  1    #    IF Y2/V4=FLOOR(Y2/V4)=0 THEN F(2)=29;
         /*GENERATE THE BIORHYTHM PLOT*/
60  1    #    L=OH+L;
61  1    #    C=0;
62  1    #    R=OH;
63  1    #    LOOP1:
64  2    #    DO OH=R TO (L-1);
65  2    #       C=C+V1;
```

/*H BIORHYTHM PLOTTING PROGRAM VERSION III */

STMT	LEV	NEST		SOURCE PROGRAM TEXT
66	1	2	#	A=' ';
67	1	2	#	A(26)='.|';
68	1	2	#	Y=0;
69	1	2	#	X=(SIN(K*(OH/V23-FLOOR(OH/V23))*V25)+V26+0.499999;
70	1	2	#	TEST1=X;
71	1	2	#	A(X)='P';
72	1	2	#	Y=Y+X;
73	1	2	#	X=(SIN(K*(OH/V33-FLOOR(OH/V33))*V25)+V26+0.499999;
74	1	2	#	TEST2=X;
75	1	2	#	A(X)='I';
76	1	2	#	Y=Y+X;
77	1	2	#	X=(SIN(K*(OH/V28-FLOOR(OH/V28))*V25)+V26+0.499999;
78	1	2	#	TEST3=X;
79	1	2	#	A(X)='E';
80	1	2	#	Y=(Y+X)/V3;
81	1	2	#	A(Y)='A';
82	1	2	#	PUT FILE(SYSPRINT) EDIT(T$(P2),M2,'/',D2,A)
				(COL(29),A,COL(34),F(2,0),COL(36),A,COL(38),F(2,0),COL(
				42),51 A);
83	1	2	#	PUT EDIT (TEST1,TEST2,TEST3) (COL(94),3 F(5));
84	1	2	#	/* INCREMENT DATE */
85	1	2	#	P2=P2+V1;
86	1	2	#	IF P2>7 THEN P2=1;
87	1	2	#	D2=D2+V1;
88	1	3	#	IF D2>F(M2) THEN DO;
90	1	3	#	D2=V1;
91	1	3	#	M2=M2+V1;
92	1	2	#	END;

```
 93   1   *      IF M2>12 THEN DO;
 95   1   3  *        M2=V1;
 96   1   3  *        Y2=Y2+V1;
 97   1   2  *      END;
 98   1   2  *      IF Y2/V4-FLOOR(Y2/V4)¬=0 THEN F(2)=28;
 99   1   2  *      ELSE F(2)=29;
100   1   1  *    END LOOP1;
101   1   1         /*        FIND DAY OF THE WEEK         */
102   1      *    GET FILE(SYSIN) LIST(Z$);
103   2      *  END LOOP3;
104   2      *  DAY: PROC;
105   2      *    N1=M2+V12*FLOOR(V6+V1/M2);
106   2      *    N2=Y2-FLOOR(V6+V1/M2);
107   2      *    N3=FLOOR(V13*(N1+V1)/V5);
108   2      *    N4=FLOOR(V5*N2/V4);
109   2      *    N5=FLOOR(N2/V100);
110   2      *    N6=FLOOR(N2/V400);
111   2      *    N7=N3+N4-N5+N6+D2-V1;
112   2      *    N7=MOD(N7,7)+V1;
113   2      *    PUT SKIP LIST ('** DAY OF WEEK CALCULATION **');
114   2      *    PUT SKIP DATA (N1,N2,N3,N4,N5,N6,N7);
115   2      *    PUT FILE(SYSPRINT) EDIT('N7= ',N7)
116   1      *                      (COL(100),A,COL(108),F(6,3));
117   2      *  END DAY;
118   2      *  JULIAN: PROC;
119   2      *    J2=0;
120   2   1  *  LOOP2:
121   2   1  *    DO I=1 TO (X-1);
122   2      *      J2=J2+F(I);
123   1      *    END LOOP2;
124   1      *  END JULIAN;
          *  END BIO;
```

Figure A.2: Printed output of biorhythm plotting program

```
BIORYTHM CHART FOR    BARBARA STOCKWELL
   BORN ON MON  7/11/1955
   BEGINNING MON 7/11/1955

   P=PHYSICAL     (23 DAYS)
   E=EMOTIONAL    (28 DAYS)
   I=INTELLECTUAL (33 DAYS)
   A=OVERALL AVERAGE
```

```
              DOWN           CRITICAL                   UP
              ----------------|------------------------------
MON  7/11                     A    IAP                        26  26  26
TUE  7/12                     |         IAP                   33  31  32
WED  7/13                     |            IAP                39  35  37
THU  7/14                     |               I A P           44  40  42
FRI  7/15                     |                I  AEP         48  43  46
SAT  7/16                     |                    IAP        50  46  49
SUN  7/17                     |                        AE     51  49  50
MON  7/18                     |                    P   AEI    50  50  51
TUE  7/19                     |                P      A E I   46  51  50
WED  7/20                     |            P       AE    I    42  51  49
THU  7/21                     |        P        AE        I   36  50  46
FRI  7/22                     |    P         AE               29  48  42
SAT  7/23                  P  |           A E        I        23  45  37
```

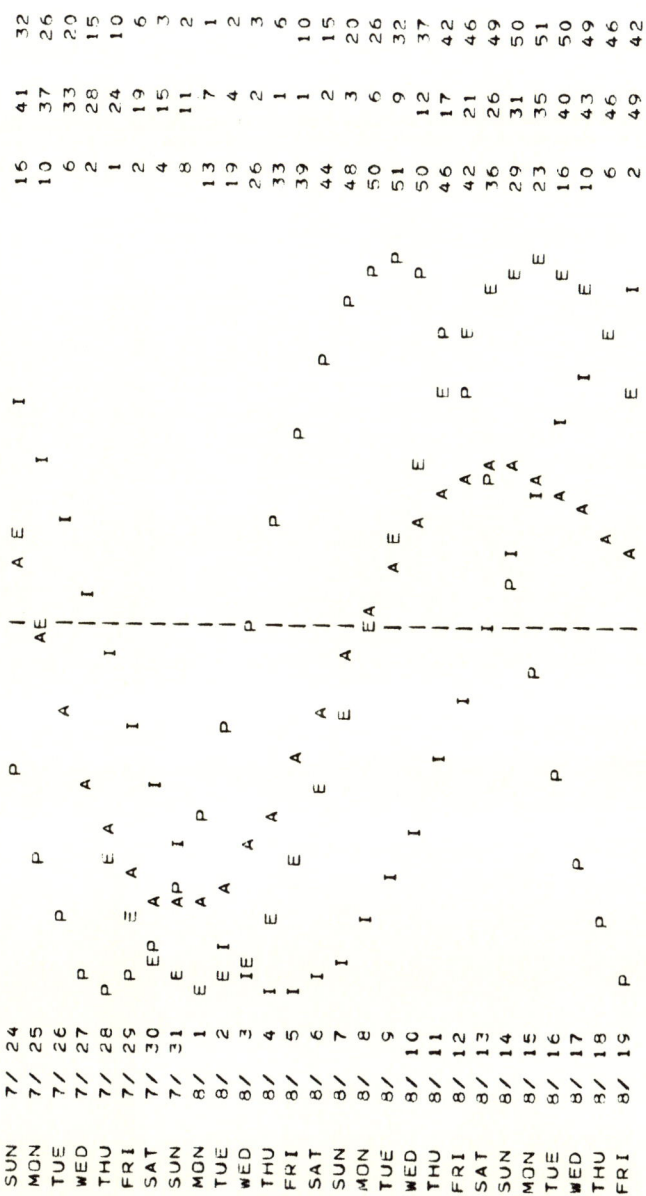

GLOSSARY

Address. An identification for a register, location in storage, or other data source or destination; the identification may be a name, label, or number.

Algorithm. The sequence of logical steps necessary to accomplish a desired goal. (A recipe is an algorithm to bake a cake.)

Alignment. The storing of data items in relation to certain machine-dependent boundaries.

Array. A named, ordered collection of data elements, all of which have identical attributes. An array has dimensions specified by a variable-defining instruction, and its individual elements are referred to by subscripts. An array can also be an ordered collection of identical structures.

Bottom-up design. The process of creating a computerized system by first starting at the lower elementary levels and then integrating higher levels until the total system is completed. This represents the traditional approach to design.

Chief programmer team. A new form of program development organization in which one person becomes basically responsible for all programming. All other team members support the chief programmer.

Complex data. Arithmetic data which has two component parts each of which consists of a real part and an imaginary part. Used primarily in scientific-oriented situations.

Composite design. A program design technology that has the aim of structuring a program into a hierarchy of highly independent logic modules.

Concatenation. To join together or connect data sets, modules, character, or bit string data.

Correct program. A logical procedure for a stated process which satisfies design specifications.

Coupling. A measure of module independence.

Data. Representation of information or of value in a form suitable for processing. Data types consist of arithmetic, string, or address.

Data aggregate. A logical collection of two or more data items that can be referred to either collectively or individually; an array or structure.

Glossary

Data base. The collection of data elements both internal and external used by a program.

Data coupling. The attribute of modules which only pass results from one to the other. This implies that such modules can be easily decoupled with only minor interface changes.

Debugging. The process of error removal from a design or program.

Decomposition. The process of taking a large set of program specifications and developing smaller, more concrete subsets.

Dimension attribute. The storage attribute that defines a storage-element as an array. In PL/I the dimension attribute consists of one or more bound pair lists to define the dimensions of an array.

Direct access. Retrieval or storage of data by a reference to its location on a disk storage device, rather than relative to the previously retrieved or stored data.

Driver program. A dummy program used to test a module of a large program by simulating its operating environment.

Dummy stub. A valid set of code which performs no actual function. It is used to test a higher level logic segment and will eventually be replaced by working code.

Fixed point. A form of arithmetic data storage where data is maintained in binary form with a fixed decimal point.

Floating point. A form of arithmetic data storage where values are maintained in the form of scientific notation (e.g., 3.14519×10^3).

Function. A process that accepts one or more inputs and produces one output. An action upon an object.

Functional decomposition. Same as iterative refinement.

Functional strength. Grouping of all logical steps to perform a single function.

Hashing. A method of converting a record key into an address which is a subset of the possible addresses defined by the actual value of the key. Hashing is a file access technique used to reduce the storage requirements of a direct access file and reduce time required to retrieve selected items.

HIPO. An acronym for Hierarchy plus Input-Process-Output. This methodology provides a graphical description of the functions performed by a system, plus the relationships among inputs, processes, and outputs of that system.

Internal procedure. A procedure which is not capable of being independently compiled, or not capable of being called from every point within a program. PL/I procedure and COBOL paragraphs are examples of this organization.

Inverted file. A file access technique used in direct file processing. Secondary key fields are maintained to improve retrieval search times. A directory of record keys associated with each inverted field value is stored and complex retrievals can be quickly accomplished.

Glossary

Iterative refinement. The evolutionary process of system design where each succeeding version has increased levels of detail. This process migrates in a top-down fashion.

Jackson methodology. A contemporary systems design methodology invented by Michael Jackson in which data structure is the key to internal logic processes.

Key word. An identifier that, when used in the proper context, has either a language-defined or an implementation-defined meaning in the program.

Linked list. A file access technique that "chains" individual, related records together by using pointers physically located in each record to define the location of the "next" element of the "chain."

Maintainability. The attribute of a program regarding ease with which it can be corrected or modified.

Meta-stepwise refinement. A design technique authored by Henry Ledgard which emphasizes stepwise decisions in a top-down manner to achieve a desired system goal.

Module coupling. A measure of data relationships among modules. It involves the mechanisms used to transmit data from one module to another.

Module strength. Logical relationships within a module all perform similar functions (e.g., all editing functions).

Parsimony (principle). The simplest solution with minimum content and variety.

Picture specification. A character-by-character description of the composition and characteristics of binary picture data, decimal picture data, and character-string picture data.

Picture specification character. Any of the characters that can be used in picture data, character-string picture data, and decimal picture data.

Pictured storage. A variable string storage type that uses the picture specifications to further describe the storage attributes.

Program design language. A pseudo-coding approach to program design developed by IBM. This is the predecessor to structured design language (SDL).

Precision. The value range of an arithmetic variable expressed as a total number of digits and, for fixed-point variables, the number of those digits assumed to appear to the right of the decimal or binary point.

Program stub. *See* Dummy stub.

Preprocessor. A technique used to modify source code before compilation. Segments can be inserted or text replaced according to programmer selected criteria.

Proper program. A program or logical segment which has precisely one entry and one exit. All structured programs should exhibit this criteria.

Pseudo-coding. An English-like expression of program logic used to bridge the gap between initial design and a high-level language implementation of the design.

Glossary

Readability. The characteristic of a program which makes individuals other than the author able to comprehend its function and process.

Record key. A key recorded in a direct access volume to identify an associated data record. *See also* direct access.

Recursive. A program, module, or subroutine that invokes itself.

Reliability. The likelihood that a program or physical device will function without failure over a given period of time.

Scalar item. A single item of data; an element.

Scalar variable. A variable that can represent only a single data item; an element variable.

Scaling. The number of digits to the right of the decimal (binary) point.

Scrutable program. A program that can be understood by humans with reasonable intelligence, provided they invest a minimal amount of effort.

Segment. An aggregation of logical instructions which produces some functional objective. Often used interchangeably with the term *module*.

Sequential access. An access mode in which logical records are obtained from, or placed into, a file in such a way that each successive access to the file refers to the next subsequent logical record in the file.

Sheltered program. One which protects the user from surprises.

Software crisis. A trait of contemporary computer systems in which the cost of software development is growing much faster than supporting hardware costs.

Stepwise refinement. An iterative process by which broad logical details are translated into firm high-level code.

Structure. A hierarchical set of names that refers to an aggregate of data items that may have different attributes (equivalent to a COBOL record).

Structured design. A set of techniques for converting a problem definition to a functional, modular program structure. Two important attributes are module coupling and module strength.

Structured design language. A formal set of English-like pseudo-code syntax which is used to translate abstract designs into formal structured code.

Structured programming. The process of converting arbitrary program logic into standard logic constructs, thereby making the resulting program easier to comprehend and modify.

Structured walk-through. A formal review session in which a particular design is evaluated by the developer's peers regarding completeness, accuracy, and general approach.

String. A connected sequence of characters or bits that is treated as a single data item.

Storage allocation. Association of a storage area with a variable.

System. A series of interrelated elements that perform some activity, function, or operation. Within a data processing environment this term usually refers to the hardware used for processing, or the set of software used to accomplish some identifiable output (e.g., the accounting system).

Glossary

Top-down design. Major functions of the system or program are designed first, followed by increasingly more detailed lower level logic. In theory, coding does not begin until the total design is completed; however, in practice this rule is seldom followed.
Tree structure. A schematic diagram used to show relationships of various sub-elements. Variable structures and program HIPO logic are two common items shown using this approach.
Warnier diagram. A contemporary schematic technique for conceptualizing a system structure.

INDEX

Accurately Designed System (ADS) 19
Address storage 181
Address types by language 182
Aggregates 176, 182
Alignment 184
Algorithm errors 153
Analysis errors 153
Arithmetic variable choices 177
Arrays 183
Attributes of good programs 143

Bohm, C. 47
Brooks, Frederick 2, 4, 201

CASE statement 4, 52, 76, 79, 85, 94
Chief Programmer Team 8
CODASYL 18
Coding standards 57, 58
Comments 88, 100, 146
Common execution time errors 154, 157, 158
Common programming errors 154, 157, 158
Compilation errors 156
Compiler error categories 155

Completion dates, estimating, 202
Composite logic components 26
Computer-assisted design techniques
 automated ADS 20
 decision table processors 20
 Time Automated Grid (TAG) 20
Condition handlers 98
Costs, hardware vs. software 14
Cougar, Daniel 15

Data structure 147
Debugging 153
 algorithm errors 153
 analysis errors 153
 case study 159
 case study solution 213
 checklist 157
 common execution-time errors 158
 common programming errors 154, 157, 158
 compilation errors 156
 compiler error categories 155
 source code errors 153
 system development errors 153
Design concepts, functional 65
Design stages
 early computer systems 16
 modern computer systems 22

Index

Dijkstra, E.W. 1, 2, 45
Direct access 92, 186
Dollar value of computers installed 13
DOUNTIL statement 47, 51, 79, 84, 93
DOWHILE statement 46, 49, 79, 84, 93

Efficiency 6, 149
ENIAC 12
External data base organization 185

Flowcharts 17, 65
Format and verifier program 104
Free-form input 148

GOTO statement 4, 65
GOTO-less programming 48

Hierarchical structuring 26
Hierarchical tree structure 34, 65
Hierarchy of components 196
HIPO charts 68
   advantages 72, 75
   detail diagram 70
   implementation 70, 74, 204
   naming conventions 75
   overview diagram (IPO) 69
   role 68
   visual table of contents (VTOC) 68
HIPO examples
   case structure 80
   conditional structure 78

iteration structure 78
sequence structure 76
Homogenous aggregate: *See* Variable specifications
Hoare, C.A.R. 45
Hughes, Joan K. 48

IFTHENELSE statement 46, 48, 78, 84, 93
Indentation 4, 85, 97, 145
Information process charts (IPC) 17
Informational algebra 18
Ingrassia, Frank 3
Input data editing 104, 148
Input/output considerations 148
Internal documentation 145
Inverted files 190
Iterative refinement 70

Jackson, M.A. 25, 26, 28, 175
Jackson methodology 25
   hierarchical structuring 26
   composite logic components 26
Jocopini, G. 47
Jones, Martha 72

Katzan, Harry 72
Kernighan, Brian 144—46

Laws of programming 150
Ledgard, Harry 27, 28
Linked list file 187
Logic design: *See* Top-down development
Logic segment 103

Index

Management control structures 196
MAP 17
Meta stepwise refinement (MSR) 27
Metzger, Phillip 204
Module types 66, 143

New York Times project 5, 72
Nonhomogenous aggregates: See Variable specifications

Off-by-one error 148

PRIDE methodology, 196, 198
Program construction 22
Program development checklist 199
Program development cycle 201
Program readability, improving 152
Program specifications, problems 62
Program standards: See Structured programming
Programmer errors 153
Programming management 197
Programming style
  comments within programs 146
  common blunders, prevention of 148
  data structure 147
  efficiency 149
  expression 144
  input/output considerations 148
  internal documentation 145
  primitive elements 151
  readability 145
  white space 146
Pseudo-coding 6, 83

See also Structured design language
Pseudo-code
  case study 112
  condition handlers 98
  data block key words 96
  design exercise 105
  guidelines 102
  key steps 102
  structure 84
  syntax 86

Readability 145

SEQUENCE statements 46, 84
Sequential access files 89, 185
Skylab project 5, 72
Software crisis 2, 3
Software design problems, traditional 42
Software as a percent of total costs 14
Source code errors 153
String storage
  ordinary 180
  pictured 180
Structured design language (SDL) 87
  comments within SDL 99
  condition handlers 98
  control structures 93
  development guidelines 103
  indentation 97
  input/output 89
  key words 90–91
  logic delineation 89
  philosophy of implementation 101
  syntax 88

Index

variable specifications 94, 96
Structured programming
  case study 124
  coding standards 4, 5, 57, 58
  constructs 1, 4, 8, 46, 48
  functional design concepts 65
  history 1, 5, 45
  language compatibility 55, 56
  module types 66, 151
  objectives 28, 143
Structured walk-throughs 8, 203
Structures 182
Study Organization Plan (SOP) 18
Synergistic effect 195
System development error sources 153
System life cycle 63

Temporary variables 145
Top-down development
  behavioral issues 42
  benefits 206
  comparison 32
  data specifications 37
  design constructs 39
  documentation 7
  implementation 8, 204
  input/output requirements 36
  logic design 1, 5, 7, 35, 38, 152
Trends in computer units installed 12

Uninitialized constant error 148

Van Tassel, Dennie 145, 149
Variable names 147

Variable specifications 94
  address storage 181
  address types by language 182
  aggregate types 176, 182
  alignment 184
  arithmetic variable choices 177
  arrays 183
  data type 176
  ordinary string storage 180
  pictured string storage 178
  pictured storage overhead 180
  structures 182
  variable types 176
  varying string storage 181

Walk-throughs: *See* Structured walk-throughs
Warnier diagram: *See* Program construction
Warnier, Jean Dominique 22, 28
Weinberg, Gerald M. 144
White space 146

Yourdon, Edward 151, 206